Dual Models

Dual Models

Magnus J. Wenninger

Cambridge University Press
Cambridge
London New York New Rochelle
Melbourne Sydney

PUBLISHED BY THE PRESS SYNDICATE OF THE UNIVERSITY OF CAMBRIDGE
The Pitt Building, Trumpington Street, Cambridge, United Kingdom

CAMBRIDGE UNIVERSITY PRESS
The Edinburgh Building, Cambridge CB2 2RU, UK
40 West 20th Street, New York NY 10011–4211, USA
477 Williamstown Road, Port Melbourne, VIC 3207, Australia
Ruiz de Alarcón 13, 28014 Madrid, Spain
Dock House, The Waterfront, Cape Town 8001, South Africa

http://www.cambridge.org

First published 1983
First paperback edition 2003

A catalogue record for this book is available from the British Library

Library of Congress Cataloguing in Publication Data
Wenninger, Magnus J.
Dual models
Bibliography: p.
1. Polyhedra–Models. I. Title.
QA491.W383 1983 516.2′3 82-14767

ISBN 0 521 24524 9 hardback
ISBN 0 521 54325 8 paperback

to Gilbert and Hugo
in grateful appreciation

Euclid alone has looked on beauty bare.

Edna St. Vincent Millay

Contents

Foreword

Mathematics is a remarkable subject. Quite simply, it is enormous fun. Usually in mathematics one is manipulating symbols on a page, or maybe these days on a computer screen, and it can be hard to appreciate the inherent beauty of the ideas one is playing with. The subject thus acquires a reputation for difficulty.

Occasionally, though, mathematics starts to describe clearly visible things, and its beauty becomes accessible to everybody. In this book, Father Magnus Wenninger takes us along one of these accessible branches of mathematics. One only has to read his work, or better still, meet the author himself, to realise that Magnus Wenninger is an enthusiast. His life's hobby is polyhedra – three-dimensional shapes governed by precise formal rules of construction. Two-dimensional shapes have their interest, but we are bombarded all our lives with two-dimensional pictures and symbols, and such patterns have lost some degree of novelty. Four-dimensional shapes are definitely difficult to visualise and appreciate.

Three dimensions make a happy compromise. A three-dimensional shape can be constructed as an actual model (though sometimes it takes some effort to do so!) and held in one's hand. Myself, I first became seriously interested in polyhedra for just this reason. One of my teachers at Cambridge, the late Jeff Miller, had contributed academically to the subject and also had an exquisite little set of intricately folded paper models. These models were mostly of uniform polyhedra, whose construction details were first becoming widely known as a result of Magnus's first book, and I amused myself for quite some time making several of them and even making my local computer discover a special one.

But what is a ''uniform polyhedron'' or a ''regular polyhedron'' or a ''stellated polyhedron'' or any of the several other classifications? The answer is that each class of polyhedra obeys a certain set of rules, laid down in advance. For uniform polyhedra, the rules happen to be that all the vertices are equivalent and all the faces are regular polygons (equilateral triangles, squares, regular pentagons, or pentagrams, etc.).

Why, one may ask, use those particular rules? The answer to that is surprising and touches the very spirit of mathematics. We use these rules because we like what they produce. At the research frontier, mathematics is a free subject. It is more of a high art form than a science. We can do what we will in mathematics, inventing such rules as we wish and playing whatever games we like with them. Our mathematical training is then used as experience in finding which rules are productive of new results, which rules are contradictory (it is conventional to discard these), and which rules interlock with other rule-systems to give new insights into the whole subtle edifice of rules (axioms) and consequences (theorems).

Sometimes, as with arithmetic, the rules enable us to deal more effectively with natural objects and are so useful that we teach them even to young children. Sometimes, as with algebra, the rules enable us to deal with generalities of objects, and we teach them to older children. Sometimes the consequences of the rules are just pretty. So it is with polyhedra. Indeed, I know of no other branch of mathematics in which the link with aesthetics is so clear.

Now, what is a ''dual model''? Magnus will explain this to you much more clearly than I would. I shall merely remark that a dual is to its original as an octahedron is to a cube and invite you to collect your paper, scissors, and glue and read on. . .

John Skilling

March 1983
Cambridge, England

Preface

In the Epilogue of my book *Polyhedron models* I mentioned that none of the Archimedean duals had been presented and also that the stellation process described in that book for two of the regular polyhedra and for the two quasi-regular solids can be applied to any of the other Archimedean polyhedra, as well as to all their duals. In my book *Spherical models* I extended the techniques of model making to the modeling of spherical polyhedra, going thereby into a deeper presentation of the mathematical basis for polyhedral symmetry. This book, *Dual models*, now completes a significant body of knowledge with respect to polyhedral forms.

In this book I propose to follow the same style as that used in the two earlier ones, presenting models in photographs, along with line drawings, diagrams, and commentary. You will find here not simply a multiplication of geometric forms but an underlying mathematical theory that unifies and systematizes the whole set of duals of uniform polyhedra. Some of these models are not as complex as some of those in the first book. Also, the mathematical approach to geometrical forms used in the second book is brought into very practical application here. So I can assure you that the level of mathematics you will need in order to follow the details of drawing and calculation will remain at the high school or secondary level. You will need some knowledge of plane geometry and some acquaintance with geometrical constructions using ruler and compass, but the three-dimensional or solid-geometry aspects will be presented in the dress of two-dimensional or plane geometry. Furthermore, the recent easy availability of small electronic calculators has taken all the tedium out of paper-and-pencil calculations that at one time depended entirely on the slide rule or on the use of mathematical tables that involved squares and square roots and trigonometry along with logarithms. It seems to me that polyhedral shapes provide an ideal topic of investigation for secondary-level mathematics.

The convex Archimedean duals are well known from recent works now available. Some of these are given in the list of references at the end of this book. But very little, or nothing, so far as I know, has been published on the duals of nonconvex uniform polyhedra. The great polyhedronist Max Brückner was acquainted with many of these, as is evident from his work *Vielecke und Vielfläche*, published in 1900, but no recent book has taken up this topic with renewed interest. The purpose of this book is to fill this lacuna.

There exists a very close relationship between the stellation process and the generation of nonconvex uniform polyhedral duals. This relationship is presented in detail in this book. It is the underlying mathematical theory that unifies and systematizes the entire set of such duals. As such, it falls directly in line with the objective mentioned in the Epilogue of my first book: "The object of an investigator would not be to multiply forms but to arrive at the underlying mathematical theory that unifies and systematizes whole sets of polyhedral forms." Furthermore, the stellation of convex Archimedean solids, and, even more, the stellation of their duals, can lead to some fantastically beautiful and interesting shapes, very useful for decorative purposes. Beauty really does not need to have uses, but the most frequent question that people ask when they see these shapes is "What do you use them for?" The aesthetic appeal that these forms have for the mind and the imagination should be enrichment enough. These shapes are interesting simply for the inherent relational aspects they have among themselves and with other simpler shapes. You will also see here that variations of strictly Archimedean forms will enter the picture, giving some hint of possibilities for the investigation of continuous transformations of geometric forms. This latter aspect, however, falls outside the scope of this book.

Special thanks must be extended to Gilbert Fleurent for the work he has done with his electronic printing calculator, from which he has obtained numerical data to produce marvelously accurate stellation patterns for the duals of the nonconvex snub polyhedra. Without his help I could never have finished this work. My thanks also to Hugo Verheyen for the interest he has taken in this work and for the suggestions he has made on how to include dual models of the so-called hemipolyhedra. I am indebted to H. Martyn Cundy for having sent to me an outline of the geometrical relationship between the stellation process and duality. This I have incorporated into the Introduction of this book. Norman W. Johnson, who was originally instrumental in ascribing names to all the uniform polyhedra, has again provided names for all the dual solids displayed in this book. Greek derivatives for the numbers and the shapes of the polyhedral parts continue to be used as heretofore, resulting in what most people find to be unpronounceably odd names. Because these names are so long and cumbersome to use, they will generally be omitted in referring to the models. The models will be designated by their numbers, the same numbers as those used in *Polyhedron models*. You will therefore find that book an indispensable companion volume to this one.

For the photography, I am indebted to Stanley Toogood, Andrew Aitken, John Dominik, and Hugh Witzmann. Excellent as the photography may be, it cannot do full justice to the models. Only when you handle a model yourself will you see the wonders that lie hidden in this world of geometrical beauty and symmetry.

M.J.W.

Introduction

The stellation process was described in some detail in *Polyhedron models*; so I shall simply refer you to that source for more detailed information. But I must say at once that a thorough acquaintance with this process is really a necessity if you want to acquire a deeper appreciation of its usefulness in relation to polyhedral duality. This will become very evident as you work your way through the models presented here.

Basic notions about stellation and duality

Very little has been published about stellations of Archimedean solids. My own work has led me to see that there is no reason to limit oneself to strictly Archimedean forms. In fact, some variations lead to far more aesthetically pleasing results, and such variations become necessary, unavoidably so, in the process of working out some of the dual forms of nonconvex uniform polyhedra. Interesting as the stellation process may be in its own way, it will not be the primary concern here.

As for duality, it seems strange in some ways that historically its geometrical significance was not clearly recognized until modern times. The five regular solids were known in ancient times, as was the entire set of thirteen semiregular solids. Johannes Kepler, in 1611, seems to have been the first to have recognized that the rhombic dodecahedron is the dual of the cuboctahedron. Other duals seem not to have entered into historical perspective until the work of E. Catalan, a French mathematician, who published his results in 1862. At the beginning of the twentieth century, M. Brückner summarized all the results of polyhedral research known at that time. In his classic work *Vielecke und Vielfläche* he gives an exact definition of duality, or what more strictly must be called the polar reciprocal relationship, and he exhibited many dual forms as models in his photographic plates.

What precisely is involved here? By way of a general description it may be said that the dual of any polyhedron is one that has the same number of edges as the original from which it is derived, but there is an interchange in the numbers of faces and vertices. The kinds of faces and vertices are such, however, that an n-sided polygon as face in the original yields an n-edged vertex in the dual. However, this is at best a rather vague definition or description of the duality relationship. Another way for you to picture the process by which the dual is generated is to fix your attention on the point called the incenter of a polyhedral face and then think of this point moving out from its position on the surface of the original polyhedron. The movement of this point (mathematically speaking, its translation) must take place along an axis of central symmetry of the solid. Such a translation will eventually bring the point to a position that coincides with a vertex point of the dual of the original polyhedron. But how far must this point move? The answer is given in the polar reciprocal relationship, which will be considered next.

Polar reciprocation

It will be useful first to introduce the notion of polar reciprocation in two-dimensional space, namely in plane geometry. In higher geometries, such as projective geometry, the notion of duality has some far-reaching consequences, but here only one aspect of it is needed, a very simple one indeed. Its very simplicity belies its far-reaching consequences even here. As a theorem in plane geometry, the basic idea of polar reciprocation can be found in Euclid. It is also directly related to the famous theorem of Pythagoras. Figure 1 shows a circle whose center is O, with a point P' inside it and P outside it, and Q is on the circumference of the circle. P is the polar reciprocal of P' if and only if $OP \cdot OP' = OQ^2$. If a, b, and r name the measured distances of P, P', and Q from O, then algebraically this theorem says that $ab = r^2$. The proof of the theorem is derived from the similarity of the right-angle triangles formed by joining P' to Q. Because corresponding sides of similar triangles are proportional, it follows that $a : r = r : b$, namely, $ab = r^2$.

In three-dimensional space (i.e., in solid geometry) you need only imagine Fig. 1 as representing a cross section through a sphere with P' inside and P outside the sphere, with Q lying on its surface and O being the center of the sphere, so that OQ becomes a radius of the sphere. The algebraic formula $ab = r^2$ can now be used to determine the exact distance of P as a vertex point of a dual in relation to the point P' taken here to represent the incenter of a given polyhedral face. This definition is very important, because the dual form can take on some very subtle transformations or variations if it is disregarded. For example, in the Epilogue of *Polyhedron models* I referred to the models made from Figs. 7, 8, and 9 (shown on p. 6 of that book) as Archimedean duals, which they really are not. In *Spherical models* (p. 51) I made reference to spherical duals. The three duals just mentioned really are plane or flat models of spherical duals. These are not the same as polar reciprocal duals. As another example, it might be pointed out that in

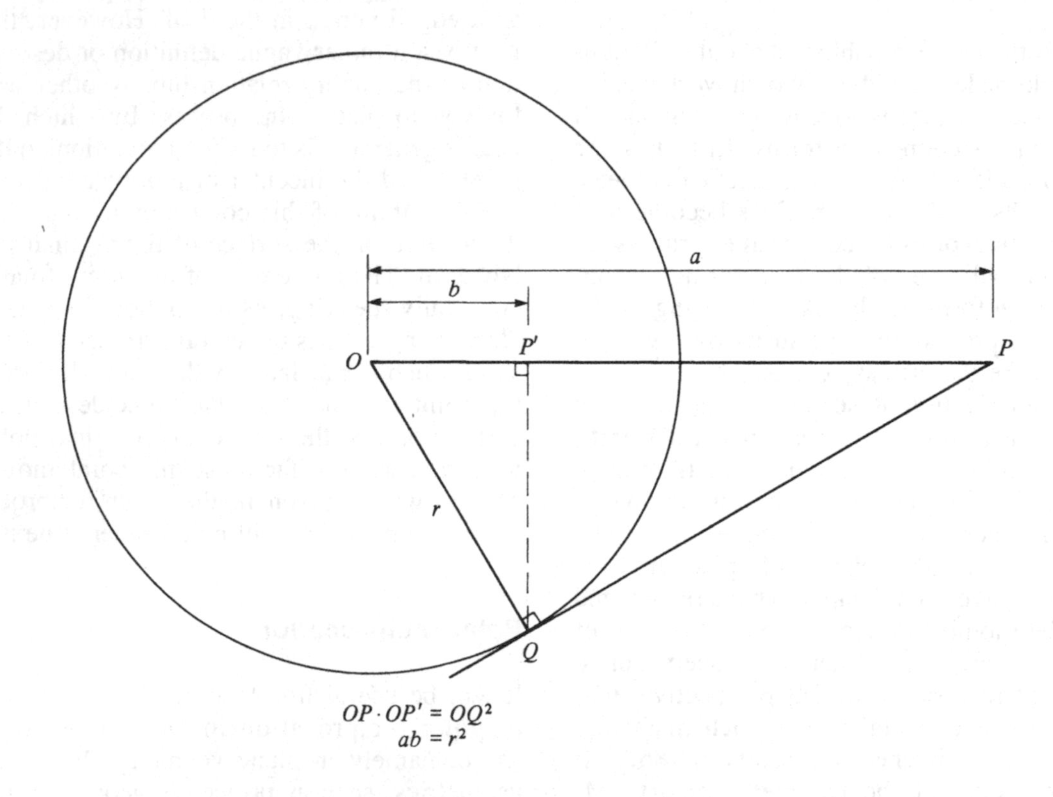

Fig. 1. Polar reciprocation in plane geometry.

a very artistically illustrated book, *Polyhedra, the realm of geometric beauty*, the author, U. Graziotti, attempted to show how the Archimedean duals can be geometrically derived by the process of erecting pyramids on the faces of a chosen basic polyhedron, the vertices of these pyramids thus determining the vertices of the related dual. Graziotti fixed his attention exclusively on the shape of the lateral faces of such pyramids and in the process neglected to consider the exact heights of such pyramids. Actual calculation involving the polar reciprocal formula disqualifies seven of the thirteen semiregular duals he intended to show.

It might be good at this time to give a fuller exposition of the polar reciprocal relationship. If you find the following presentation too abstract on a first reading, you may want to skip it for now and return to it later on, when the handling of models may greatly aid you toward a better understanding. I am indebted to H. Martyn Cundy for the following elaboration. He sent this to me after reading an article I wrote in which I enunciated the following conjecture: "The dual of any given non-convex uniform polyhedron is a stellated form of the dual of the convex hull of the given solid." The convex hull (which Coxeter calls the "case") is the smallest convex solid that can contain it. The dual of this convex hull is either known or can be found by using the polar reciprocal formula. Once it is found it serves as the "core" of the stellation process.

Here is Cundy's summary:

1. Every uniform polyhedron has all its vertices lying on a sphere.
2. The process of forming the dual is equivalent to taking the polar reciprocal in this sphere.
3. In polar reciprocation,
   every point is replaced by its polar plane
   every plane is replaced by its pole.
4. If the sphere has center O and radius r, the polar plane p of the point P is the plane normal to OP through N, where N is on OP and $OP \cdot ON = r^2$. See Fig. 2.
5. If P is on the sphere, p is a tangent plane at P. If P is outside the sphere, the tangents from P to the sphere meet it at points on p. If P is on q, the polar plane of Q, then Q is on p. See Fig. 3.
6. The plane p is between O and Q if and only if the plane q is between O and P. See Fig. 4.
7. The convex hull of a uniform polyhedron is a polyhedron whose dual is a convex polyhedron with an inscribed sphere touching all its faces.

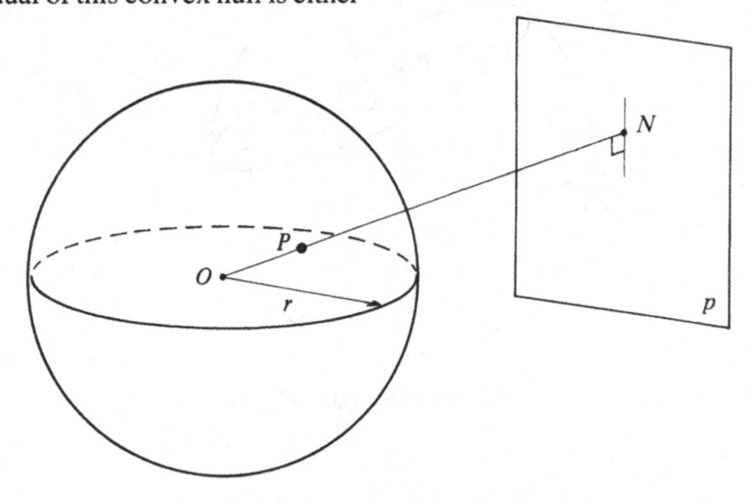

Fig. 2. Polar reciprocation in solid geometry.

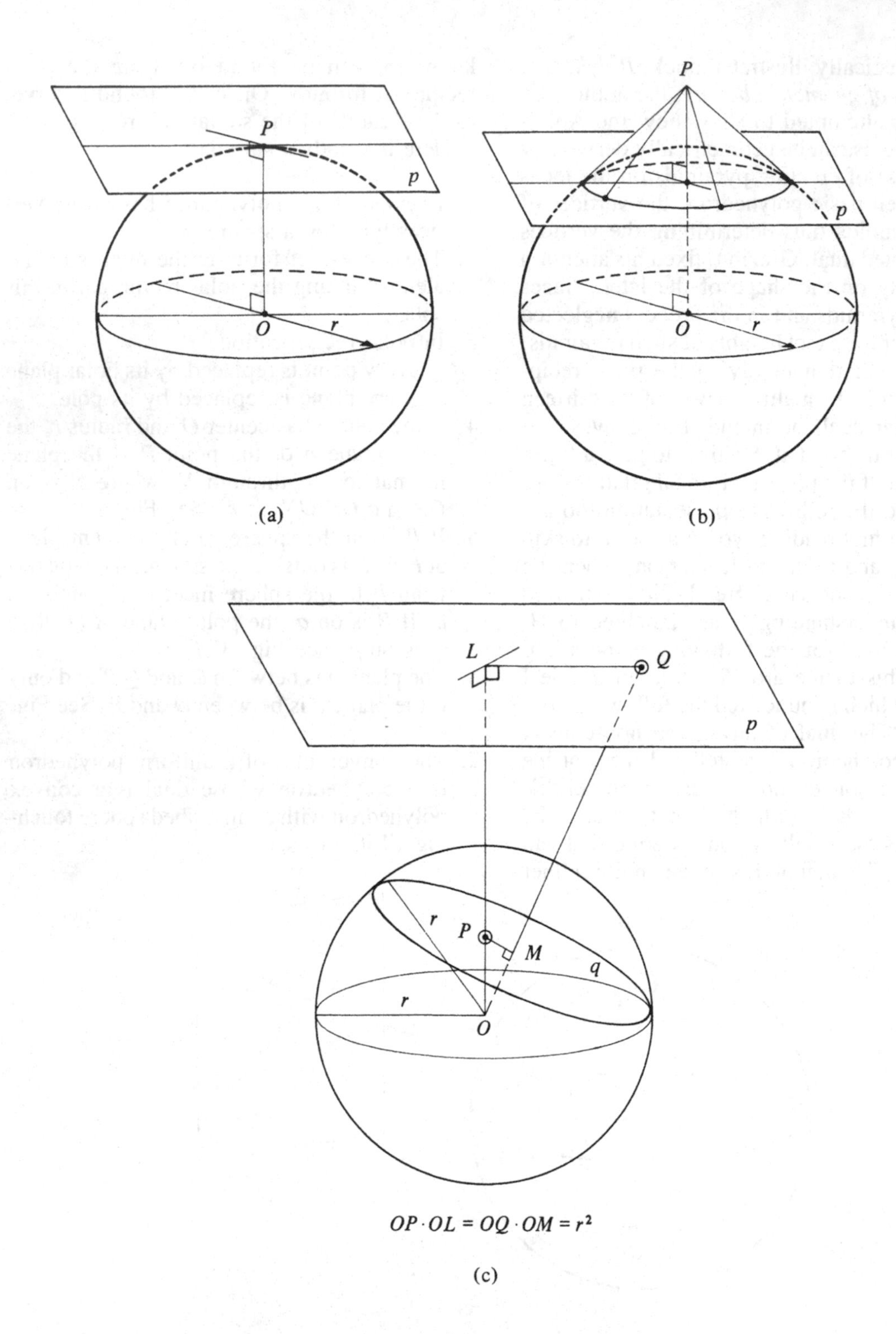

Fig. 3a–c. Polar reciprocation in solid geometry.

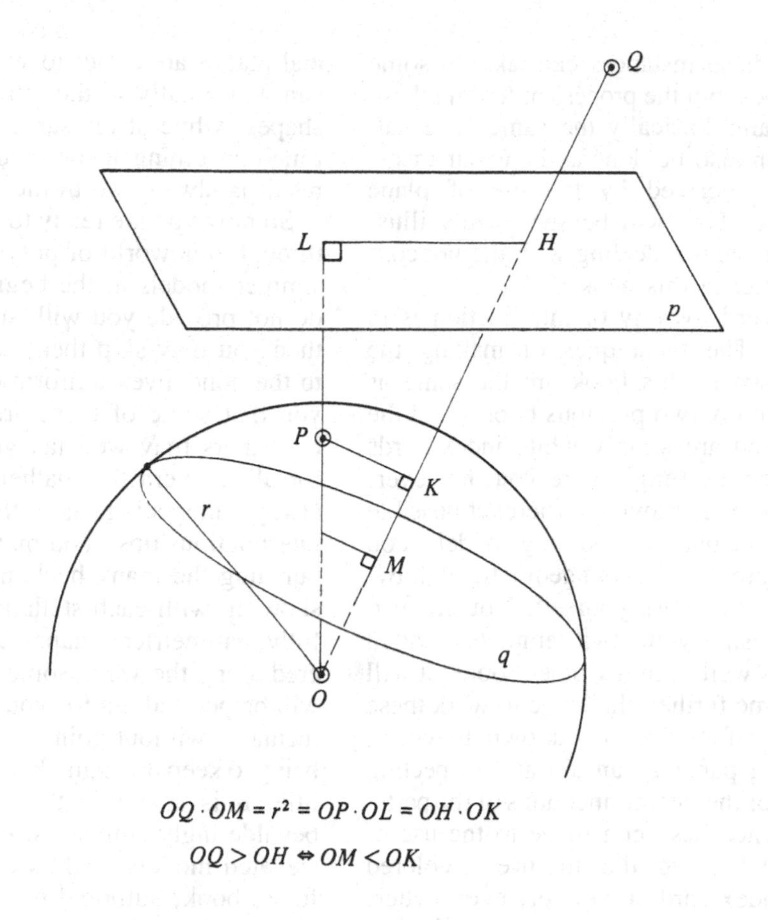

Fig. 4. Polar reciprocation in solid geometry.

8. A uniform polyhedron has all its faces either lying in the convex hull or passing between the center and some vertex of the hull. So the vertices of the dual either lie in the planes of the dual of the hull or lie outside some of its planes. Furthermore, because the vertices of the polyhedron are vertices of the hull, the planes of the dual are the planes of the dual of the hull. But this means that the dual is a stellation of the dual of the hull. Thus, all the nonconvex uniform polyhedral duals are stellations of the duals of their convex hulls.

This much from Cundy.

How facial planes are embedded in stellation patterns

With respect to the stellation process, a distinct advantage enters here because the duals have each a single stellation pattern because of the fact that a dual form is isohedral (i.e., all its faces are alike or congruent), just as the original is isogonal (i.e., all its vertices are alike). Thus, the face of the dual of any nonconvex uniform polyhedron is embedded in the stellation pattern of the dual of its convex hull.

The faces of nonconvex duals can be derived from the vertex figures of the original nonconvex forms. The polygons that appear as faces in these instances can take on some strange shapes, but the process of finding these shapes remains basically the same. The calculations can also be done and angular measures can be derived by the use of plane trigonometry. This will be specifically illustrated in the section dealing with the nonconvex duals later in this book.

A final word by way of introduction is in order here. The techniques of making the models shown in this book are the same as those used in my two previous books, and the materials used are simply white index cards or stiff paper. I strongly urge you, however, to make your own drawings wherever possible before you set out to make any model. You will learn a great deal about geometrical drawing by doing the work yourself. You are also invited to design your own templates, called *nets* here, as well as in my other books. It will give you some further challenge to work these out for yourself looking at your own drawings, using tracing paper as an aid and inspecting the photos for the interconnections of the parts.

No reference has been made to the use of color, but it is a fact that the use of colored tag (i.e., index-card stock), or, even better, metallic papers, can add greatly to the beauty of these shapes. Colors may also be introduced after a model has been constructed, when facial planes are easier to see. The use of color can add greatly to the attractiveness of these shapes, while at the same time adding to the time-consuming labor involved. But the end result is always worth the effort.

So now you are ready to begin your journey through this world of polyhedral duals. If the simpler models at the beginning of the book do not provide you with sufficient challenge, then you may skip them and go immediately to the nonconvex uniform duals. I can assure you that some of these are not too difficult, but others may well tax your ingenuity. But for all of them the mathematical aspects are always interesting, as is the richness of their interrelationships. You may also find yourself pursuing the many beckoning side paths that show up with each stellation pattern. Beautifully symmetrical shapes can thus be discovered along the way. Some of these side paths will be pointed out for you in the related commentary, without going into them, the reason being to keep the main theme true to the book's title. It is also true that stellated forms are bewilderingly numerous. A fuller treatment of stellated models could well be the topic of a future book, authored not necessarily by me, perhaps, but by anyone who finds never-ending pleasure in the discovery of new polyhedral shapes.

I. The five regular convex polyhedra and their duals

The five regular solids, also called the Platonic solids, are well known. If you have these as models to work with now, you will find that the notion of duality can very easily be illustrated with regard to them.

The tetrahedron is the simplest of all polyhedra. It has only four faces, each of which is an equilateral triangle. It has four trigonal vertices, which means that three face angles surround each vertex. Finally, it has six edges. You see immediately that an interchange in number and kind of faces and vertices leaves the number four unchanged. Because the dual of any polyhedron always keeps the same number of edges as the original from which it is derived, the number six must be kept for the number of edges. This simple description shows you that the tetrahedron is its own dual; that is, the dual of a tetrahedron is another tetrahedron.

If you look now at the octahedron, you see that it has eight faces, each of which is an equilateral triangle. It has six vertices, which can be called tetragonal, because four face angles surround each vertex. Finally, it has twelve edges. An interchange in number and kind of faces and vertices implies that its dual must have eight trigonal vertices and six tetragonal faces, and its edges must still number twelve. Another word for tetragonal is the word quadrangular, a figure or polygon with four angles. This means it also has four sides. But the question is What shape must it have? The answer is found by observing that the octahedron has a square as its vertex figure. Therefore the dual must have faces in the shape of a square. If three such squares must surround each vertex, you see that the dual of the octahedron must be the cube or hexahedron.

Now look at the cube and go through the same steps of consideration. The cube has six faces, each a perfect square. It has eight trigonal vertices and twelve edges. Its vertex figure is an equilateral triangle. So the dual of the cube must have eight equilateral triangles for faces and six tetragonal vertices, and the number of edges remains twelve. But this is a description of the octahedron. So you see that the dual of the cube is the octahedron.

Two more regular polyhedra are left to complete this introductory investigation of the five regular solids. They are the icosahedron and the dodecahedron. The icosahedron has twenty equilateral triangles for faces. It has twelve pentagonal vertices and thirty edges. An interchange in number and kind of faces and vertices, while retaining the number of edges, will lead you to see that its dual is the dodecahedron. Because the dodecahedron has twelve pentagonal faces and twenty trigonal vertices, with the number of edges still remaining at thirty, you see that its dual is the icosahedron.

This simple investigation shows that the five regular solids have duals within the same set of five. But this investigation did not take into account the polar reciprocal relationship. This can be applied mathematically, but the results are not particularly enlightening. This is so because, first of all, it is too powerful a tool to use for such simple shapes and, secondly, reciprocation in the midsphere rather than in the circumsphere gives a more interesting result. Such a reciprocation implies that the respective edges of any one of the regular solids can be made to become the perpendicular bisectors of the corresponding edges of its dual. This is a particularly attractive arrangement for model making, one that is usually shown in books about polyhedra. The tetrahedron with its dual becomes a compound of two tetrahedra, a stellated form of the octahedron, as shown in *Polyhedron models* **19**. The octahedron with its dual, the cube, is a stellated form of the cuboctahedron, as shown in *Polyhedron models* **43**. Finally, the icosahedron with its dual, the dodecahedron, is a stellated form of the icosidodecahedron, as shown in *Polyhedron models* **47**.

A different technique for model making is suggested in this book for the five regular solids. This technique can then be carried over very satisfactorily into the models of semiregular solids.

Photos 1 through 5 show each of the five regular solids embedded inside its own dual, but this dual appears only as an edge model. The geometrical and numerical details follow the elaboration given in *Spherical models* (pp. 125–31). These models are designed so that a vertex of the inner solid coincides with the incenter of a face of the dual. The faces of these duals, however, are merely suggested by the edges that lie outside the respective edges of the inner model. Thus, the edges are still perpendicular bisectors of each other, but not through each other, called skew lines; that is, the midpoint of one edge lies directly above the midpoint of the other on a radial line or a central axis of symmetry. Figures 5 through 9 show how the parts for these models may be drawn. First, a vertex part of the inner solid is laid out, but this part is shown in relation to the entire face of the inner solid. Then a drawing is given for the part needed to make an edge model of the dual. The lower-case letters on these drawings indicate where tabs are needed for cementing the parts together.

You may begin your work of making these models by first making your own templates or nets. Then cut out from card stock or stiff paper a sufficient number of parts, as many as may be needed in each instance. Next cement the vertex part into the edge design part, forming a trigonal, tetragonal, or pentagonal cup or inverted pyramid without a base holding a vertex part inside it. Finally, cement all these cups or inverted pyramids together, the lateral face of one to the lateral face of another, until the model is complete.

You may, of course, alter the design of any one of these, or all of them, to show other

Photo 1. Tetrahedron (**1**).

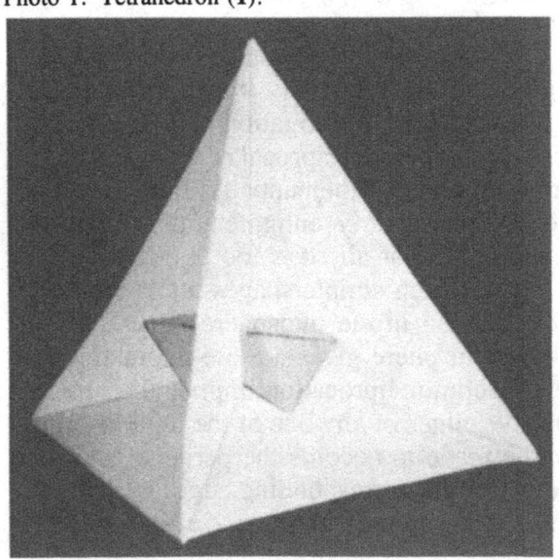

Photo 2. Cube or hexahedron (**2**).

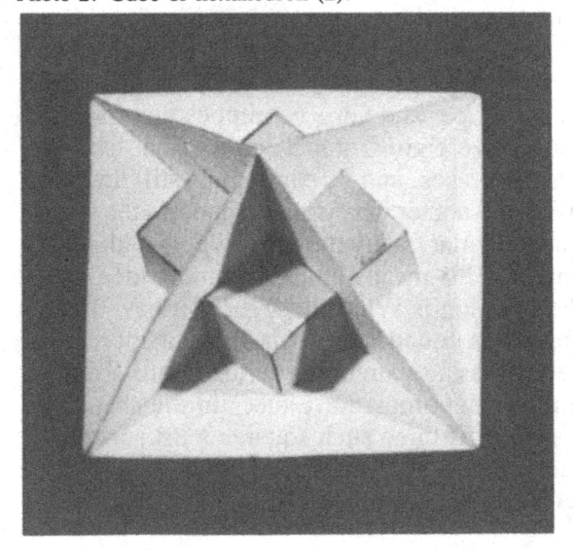

Photo 3. Octahedron (**3**).

Photo 4. Dodecahedron (**4**).

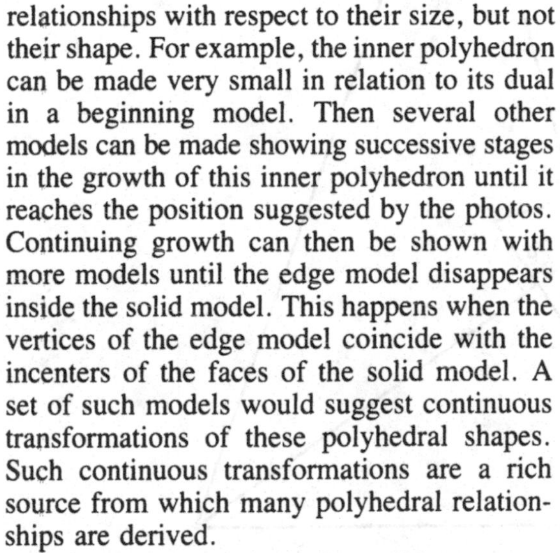

relationships with respect to their size, but not their shape. For example, the inner polyhedron can be made very small in relation to its dual in a beginning model. Then several other models can be made showing successive stages in the growth of this inner polyhedron until it reaches the position suggested by the photos. Continuing growth can then be shown with more models until the edge model disappears inside the solid model. This happens when the vertices of the edge model coincide with the incenters of the faces of the solid model. A set of such models would suggest continuous transformations of these polyhedral shapes. Such continuous transformations are a rich source from which many polyhedral relationships are derived.

Although the five regular solids are very simple shapes in themselves, within them there lies hidden the whole world of polyhedral symmetry. Like musical variations on a theme, the five regular solids reveal their presence in countless ways in all the more complex shapes that will appear later on in this book. You will see that this is already true in the next set of uniform polyhedra to be considered (i.e., the thirteen semiregular solids and their duals).

Photo 5. Icosahedron (**5**).

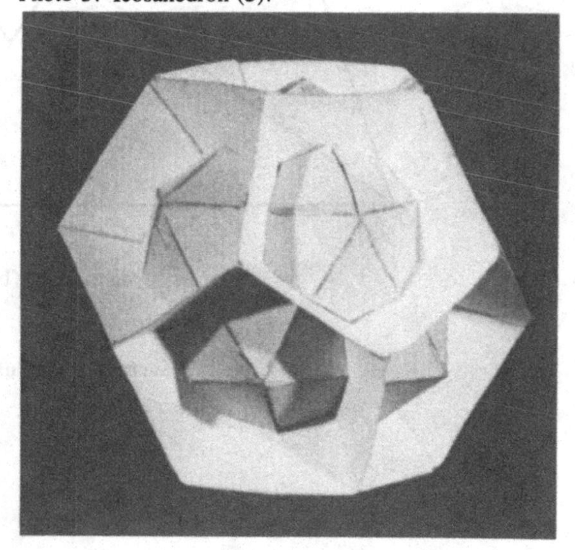

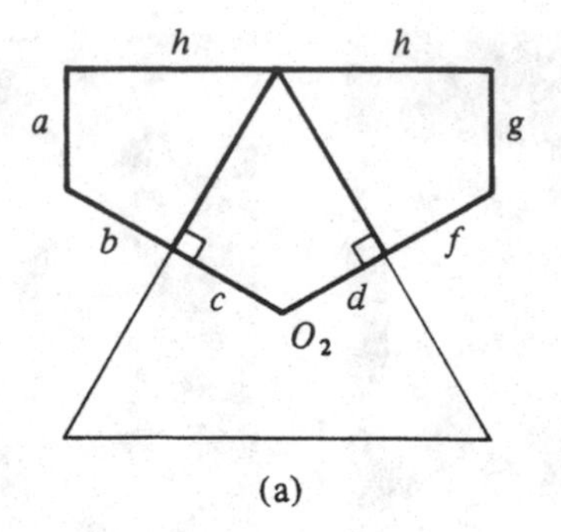

(a)

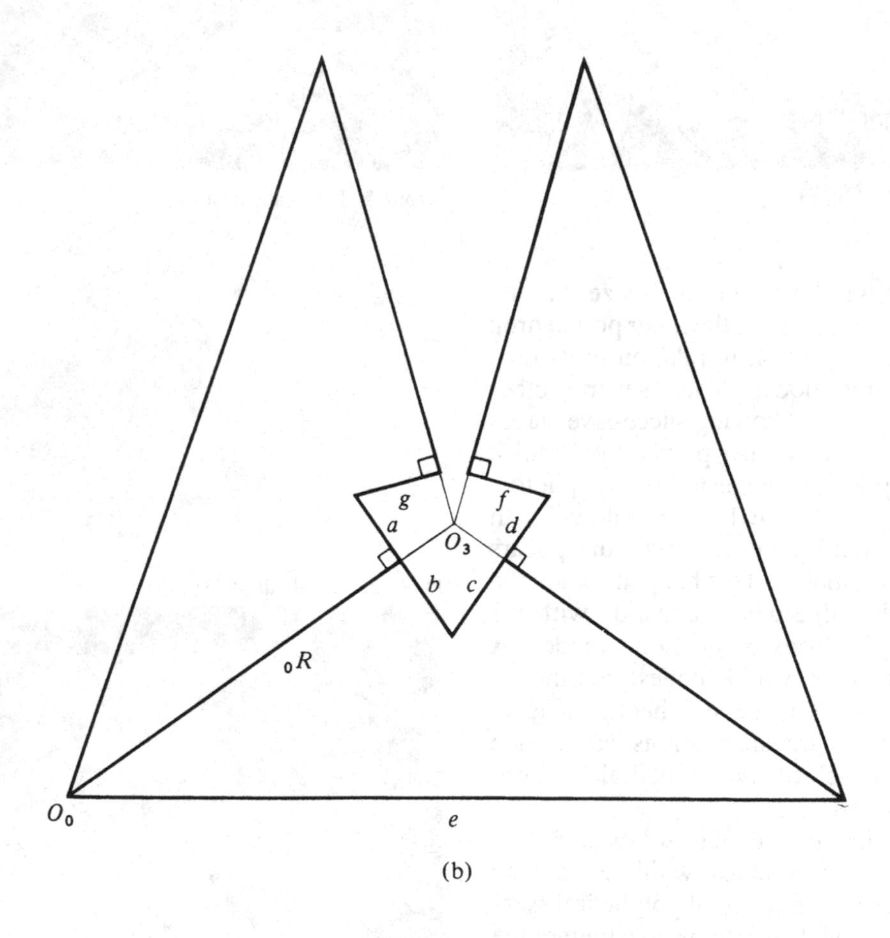

(b)

Fig. 5. Patterns for the dual of the tetrahedron (**1**).

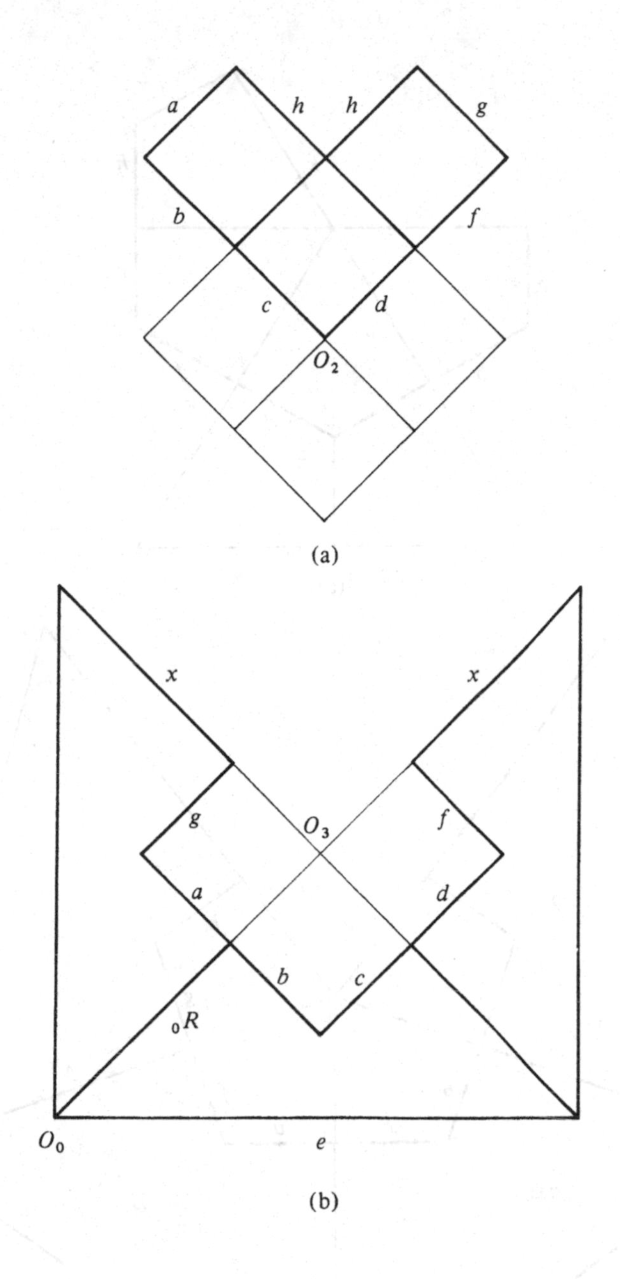

Fig. 6. Patterns for the dual of the octahedron (**2**).

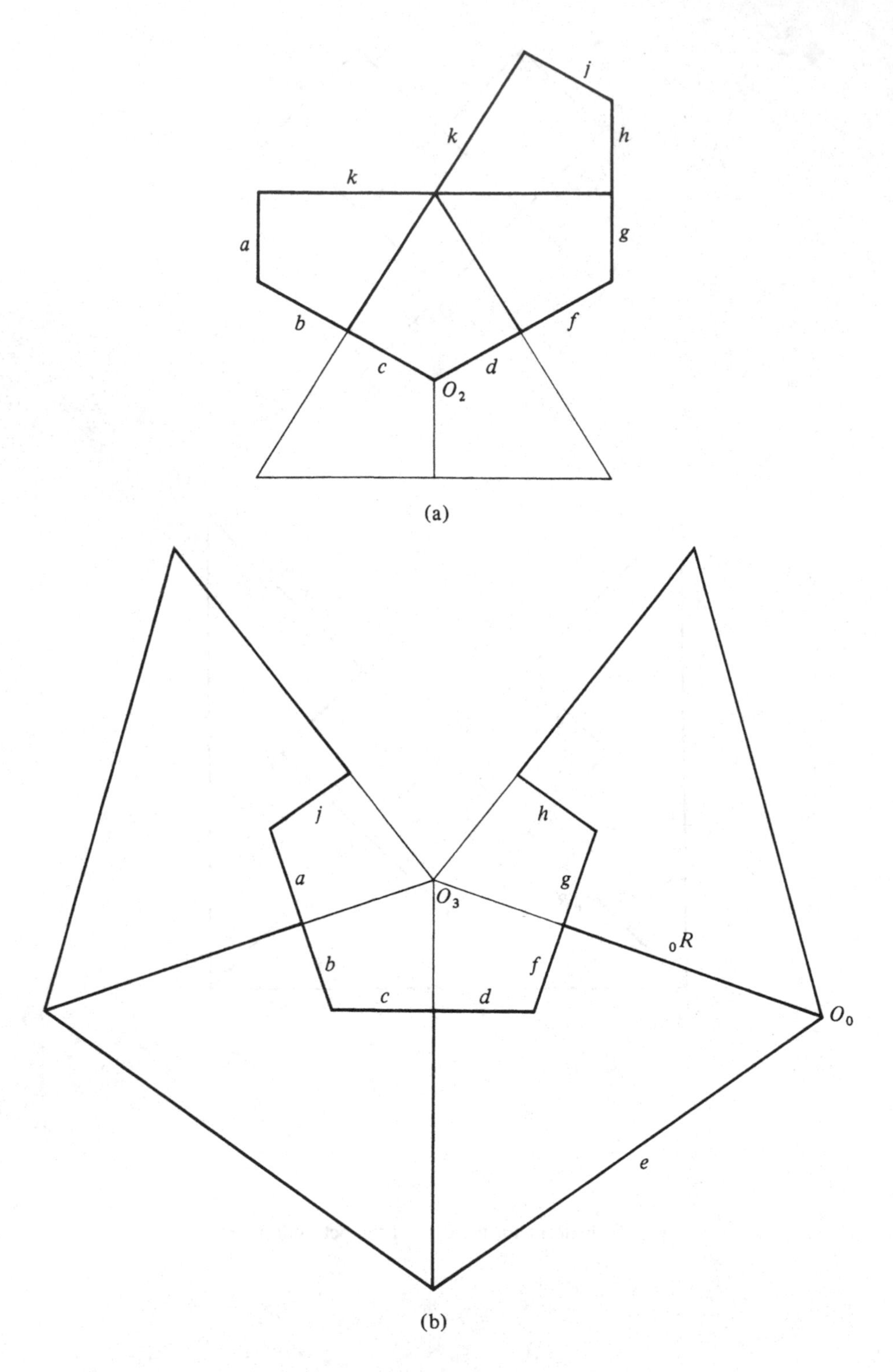

Fig. 7. Patterns for the dual of the cube (**3**).

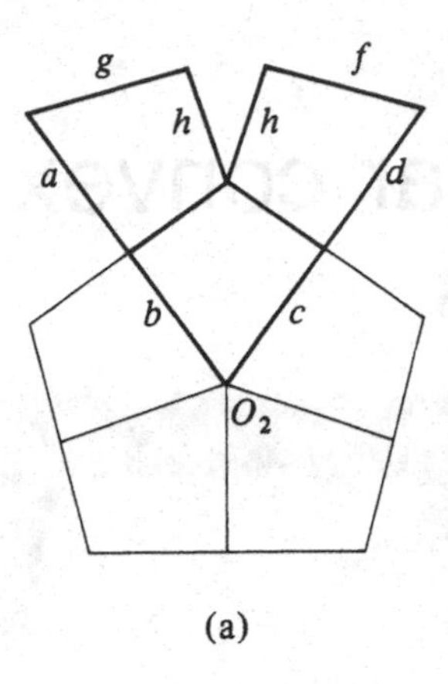

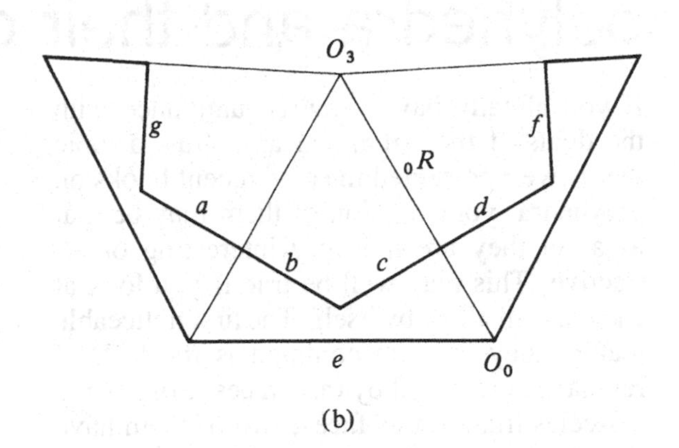

Fig. 8. Patterns for the dual of the icosahedron (**4**).

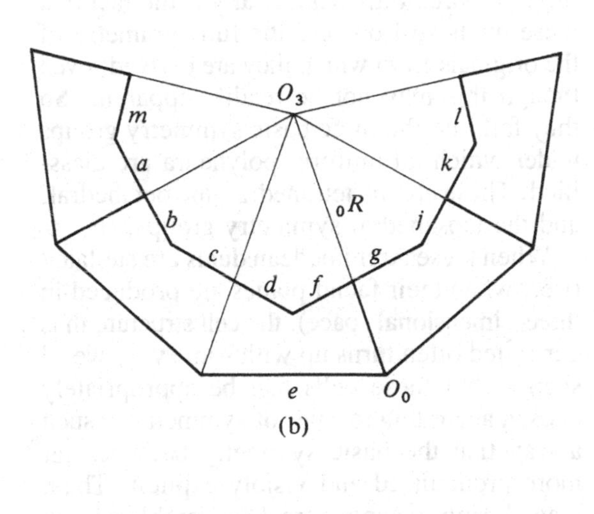

Fig. 9. Patterns for the dual of the dodecahedron (**5**).

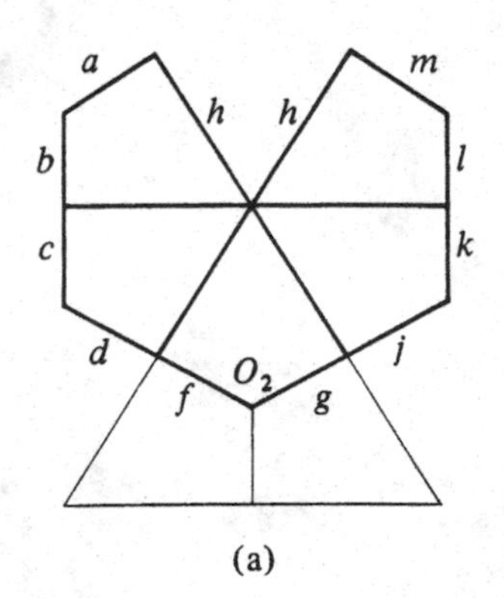

II. The thirteen semiregular convex polyhedra and their duals

If you already have some acquaintance with the duals of the Archimedean solids, a topic that is well presented in some recent books on polyhedra, your opinion of them may be that as a set they are not very interesting or attractive. This may well be true if you look at each one all alone by itself. The first noticeable feature they have in common is the lack of regularity presented by their faces. Some have isosceles triangles as faces, two of them have rhombic faces, others have kite-shaped faces designated more exactly deltoidal, and the rest have only scalene triangles for faces. Moreover, vertices can be of different kinds, following the fact that the originals have faces of different kinds, and these vertices can be at different distances from the center of symmetry when they are of different kinds. But more important mathematically is the fact that these duals still possess the full symmetry of the originals from which they are derived, even though this may not be readily apparent. So they fall into the three basic symmetry groups under which all uniform polyhedra are classified. These are the tetrahedral, the octahedral, and the icosahedral symmetry groups.

When these Archimedean duals are stellated (i.e., when their facial planes are produced in three-dimensional space), the cell structure thus generated often turns up with some very weird shapes, but these cells can be appropriately chosen according to rules of symmetry in such a way that the basic symmetry becomes far more pronounced and visibly explicit. Thus, many beautiful shapes are discoverable among the stellated forms of Archimedean duals, as well as among some variations of these duals.

Photos 6 through 18 show how models of the Archimedean duals may be embedded in their respective originals. However, in these

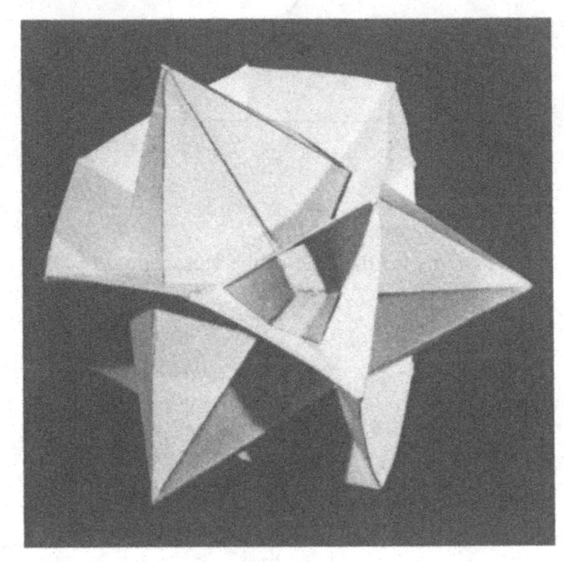

Photo 6. Triakistetrahedron (**6**).

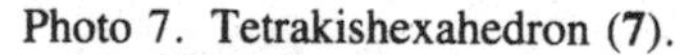

Photo 7. Tetrakishexahedron (**7**).

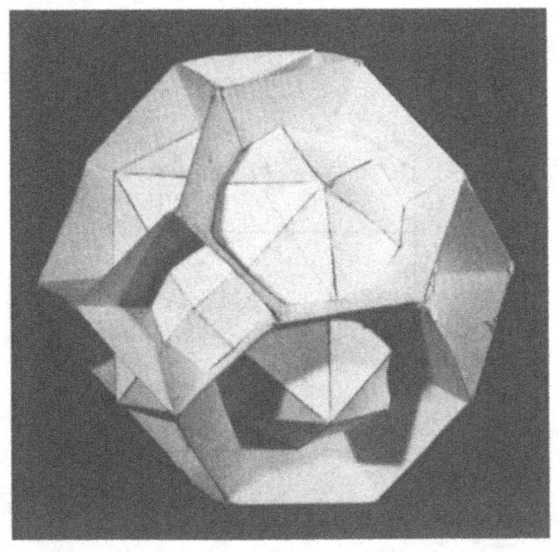

models no attempt has been made to have the vertices of the duals coincide with the incenters of respective faces. Rather, a more pleasingly aesthetic motive or simple practicality prompted the actual choice of an arbitrary scalar relationship of the one to the other. The technique of assembly for these models is the same as that suggested for the regular solids. In Figs. 10 through 22, one drawing shows one face of the dual, with lines drawn perpendicular to the sides of this face, meeting at a point called the orthocenter of that face.

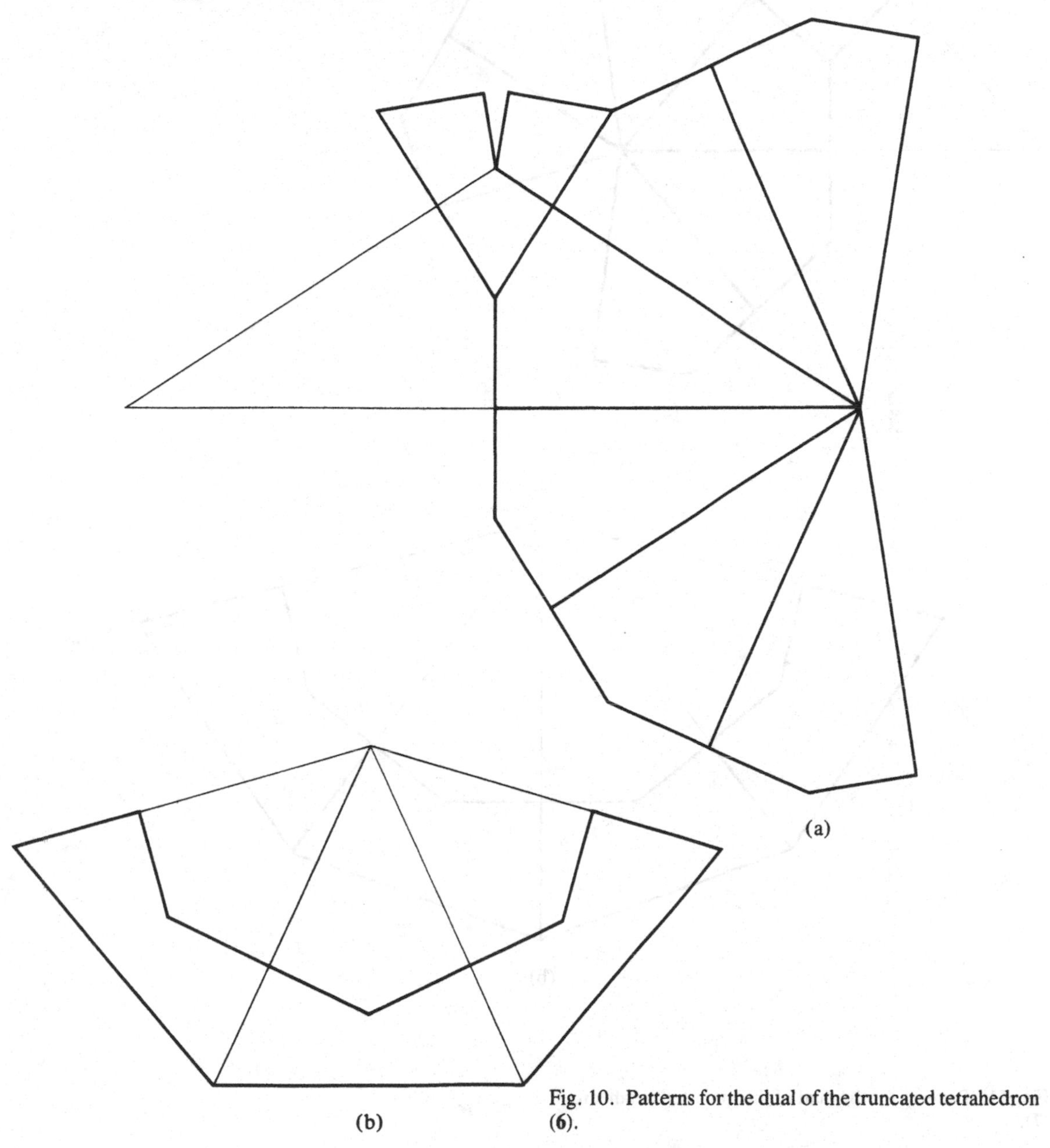

Fig. 10. Patterns for the dual of the truncated tetrahedron (6).

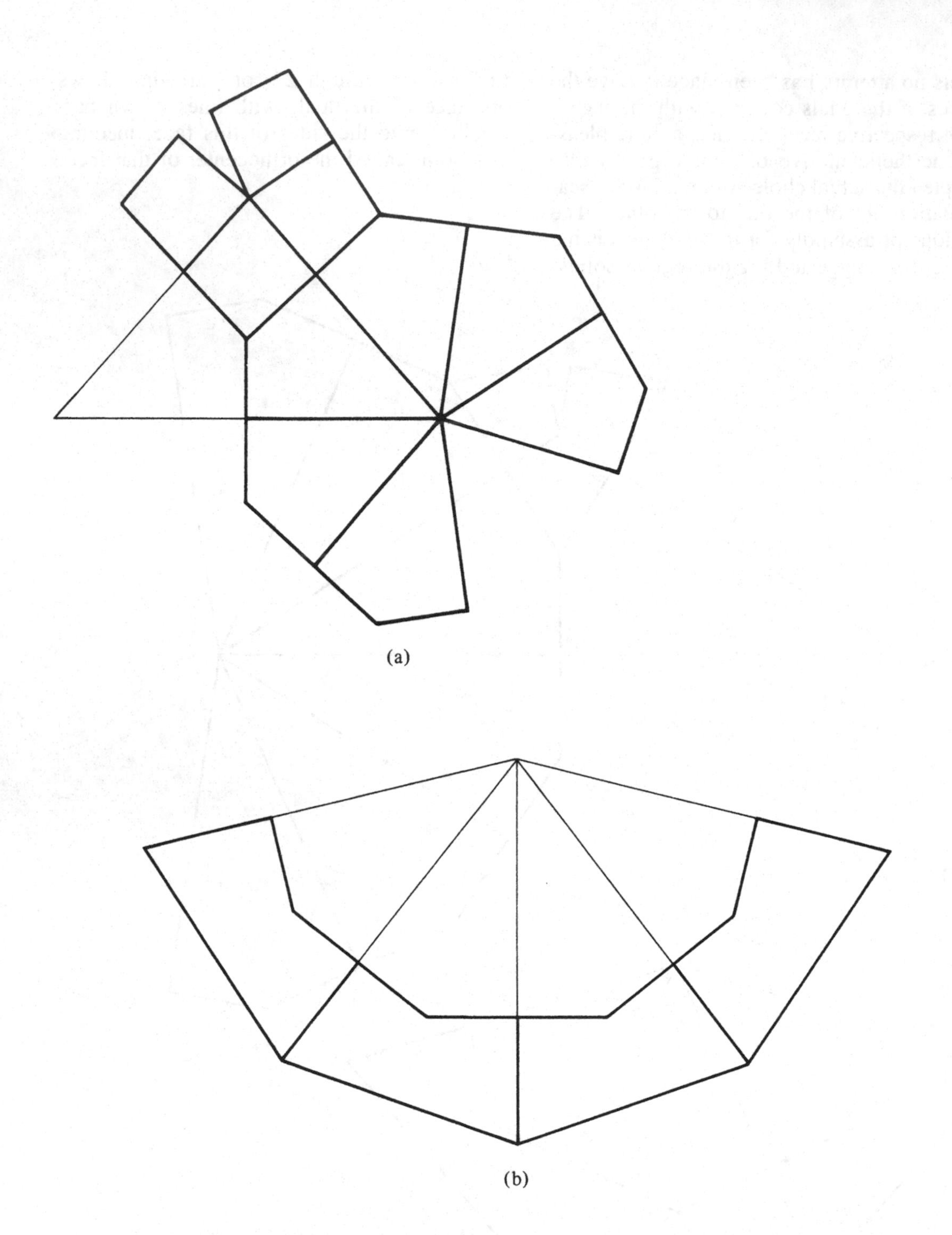

Fig. 11. Patterns for the dual of the truncated octahedron (7).

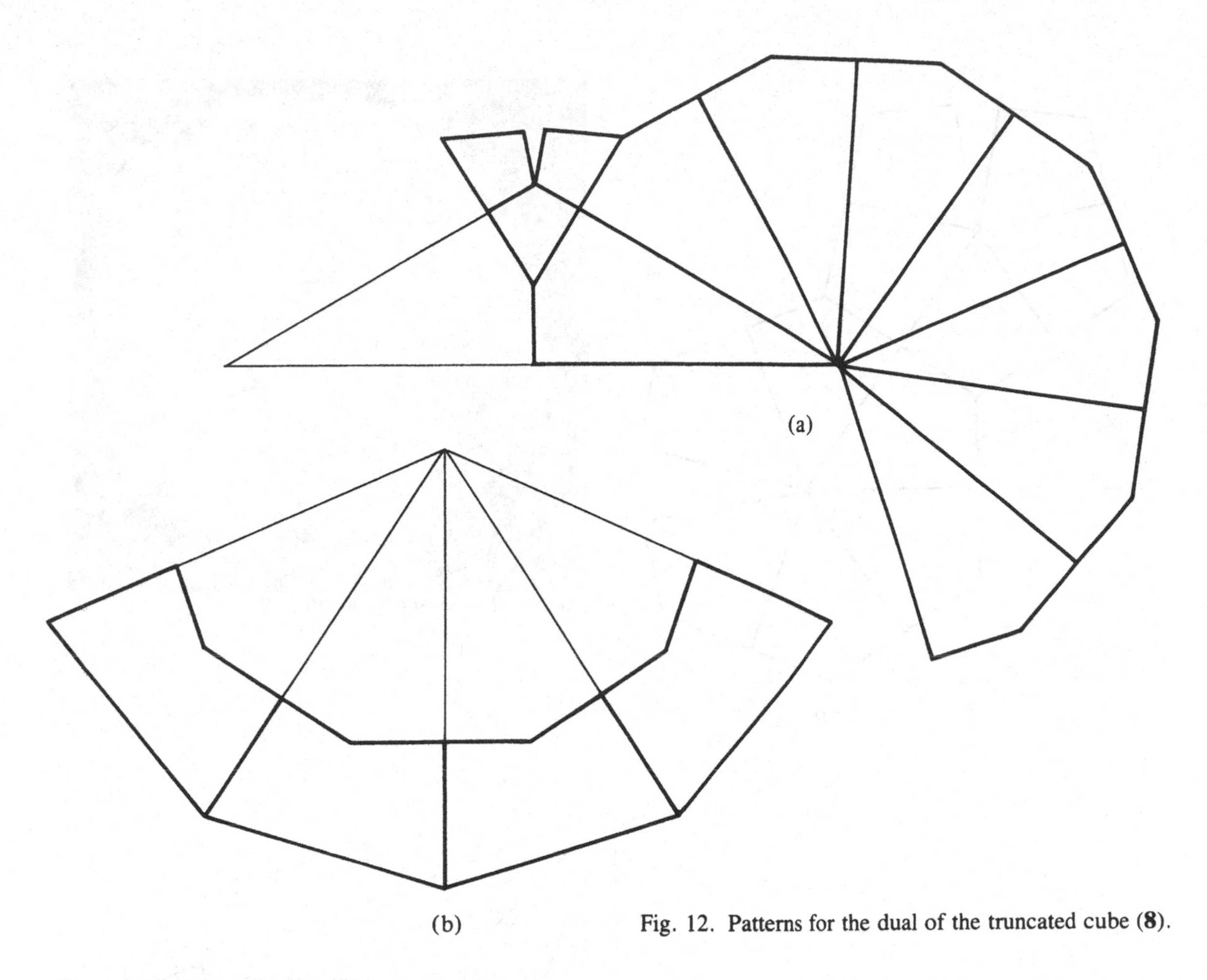

Fig. 12. Patterns for the dual of the truncated cube (**8**).

Photo 8. Triakisoctahedron (**8**).

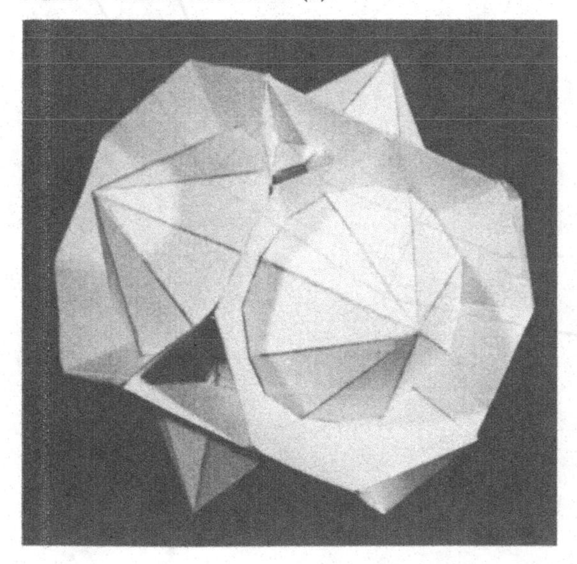

Patterns or templates are shown in connection with the face, giving vertex parts for each dual. Notice that here two or three different shapes of such vertex parts appear, in conformity with the fact that the duals always have two or three different kinds of vertices. A second drawing shows the design of the related edge model of the original. The assembly of the parts is done in the same manner as described previously for the regular solids. The photos should give you a good idea of what the end result comes out to be.

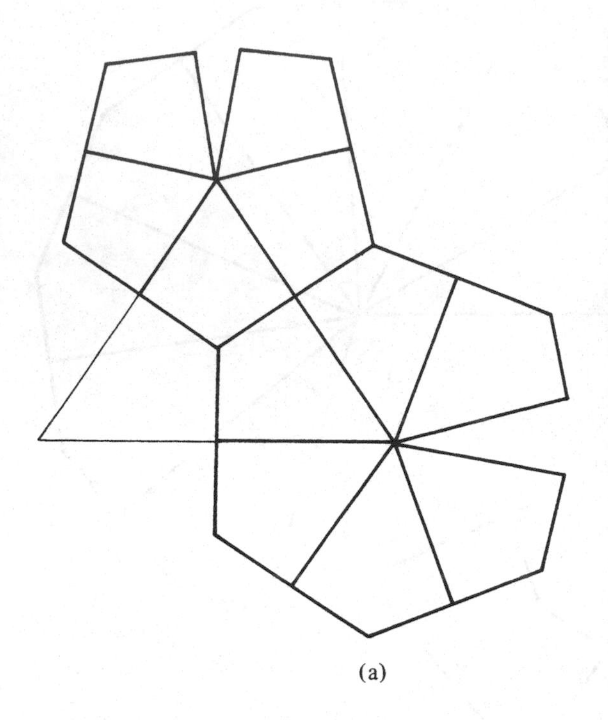

(a)

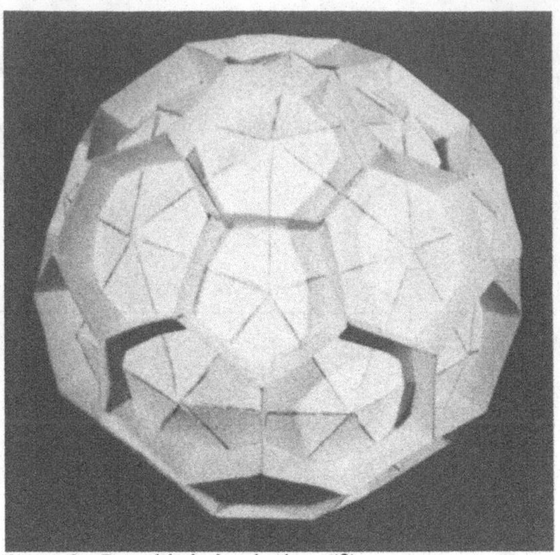

Photo 9. Pentakisdodecahedron (**9**).

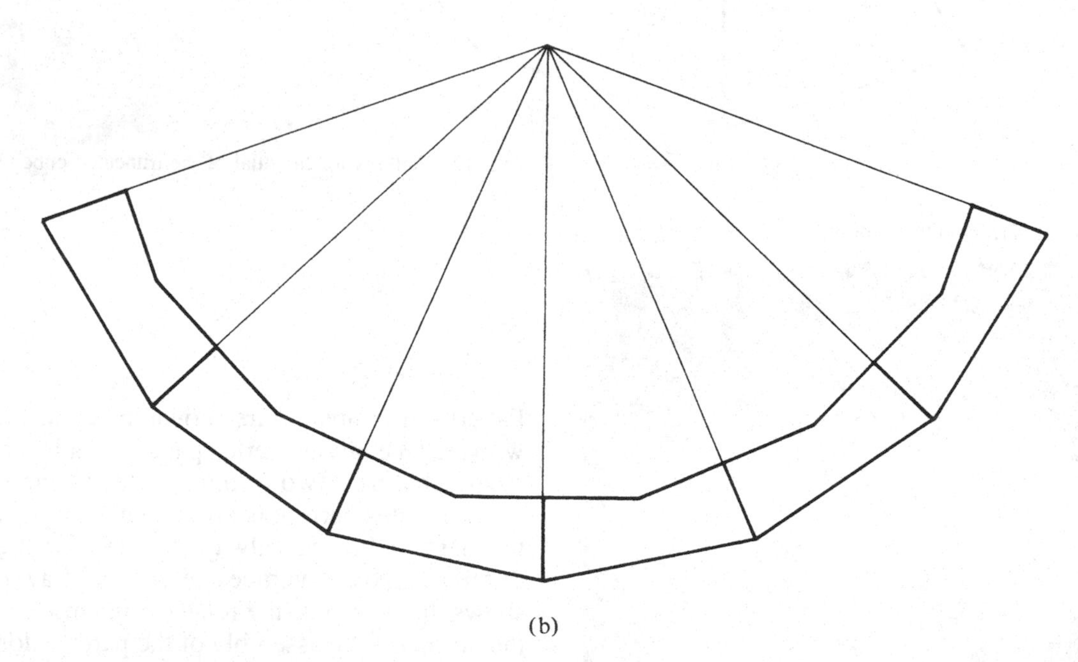

(b)

Fig. 13. Patterns for the dual of the truncated icosahedron (**9**).

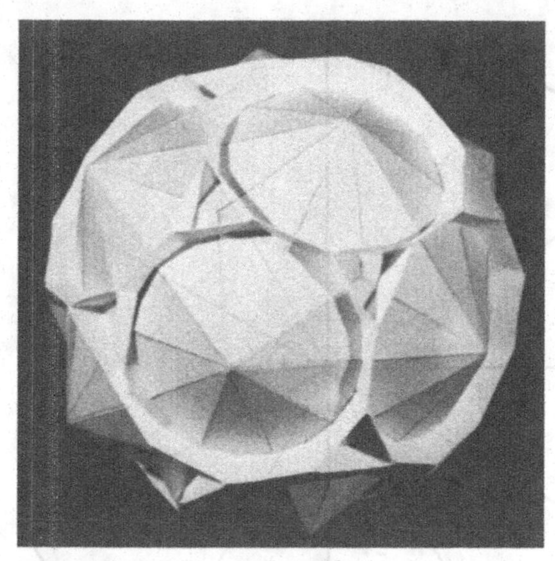

Photo 10. Triakisicosahedron (**10**).

Photo 11. Rhombic dodecahedron (**11**).

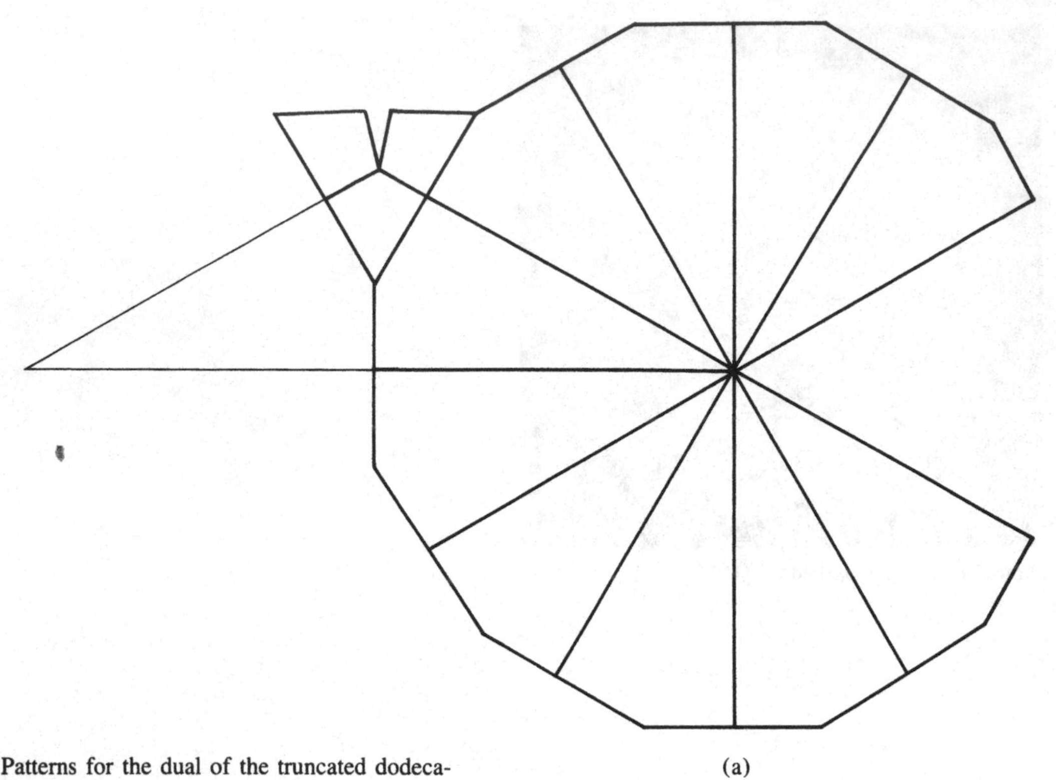
(a)

Fig. 14. Patterns for the dual of the truncated dodecahedron (**10**).

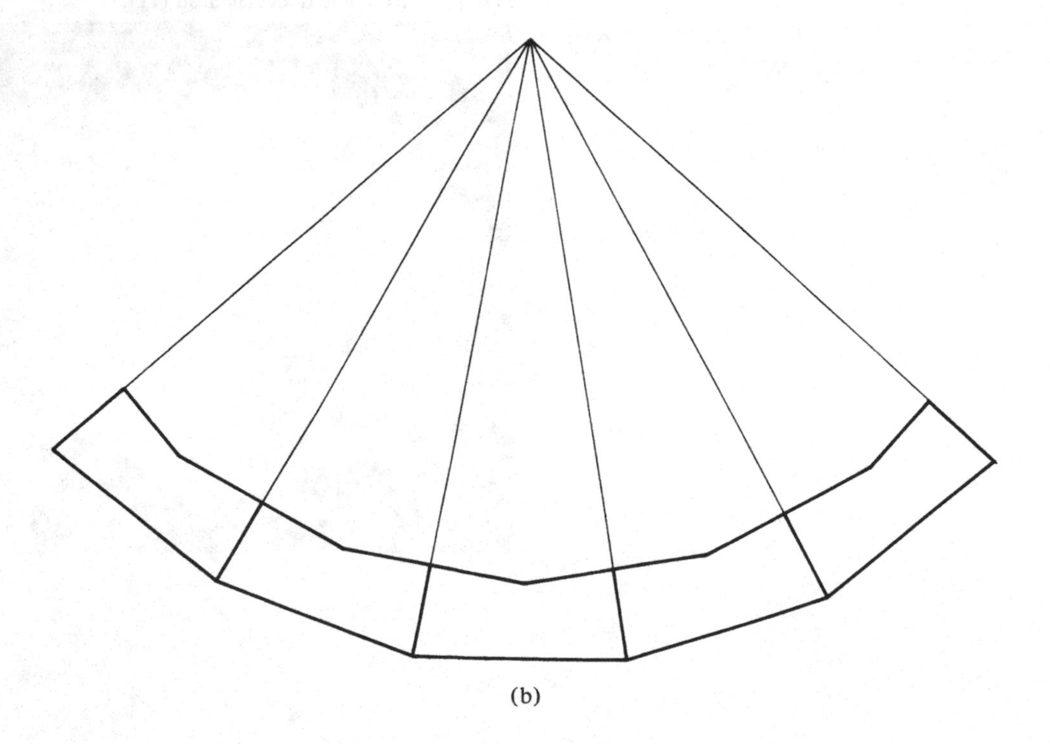
(b)

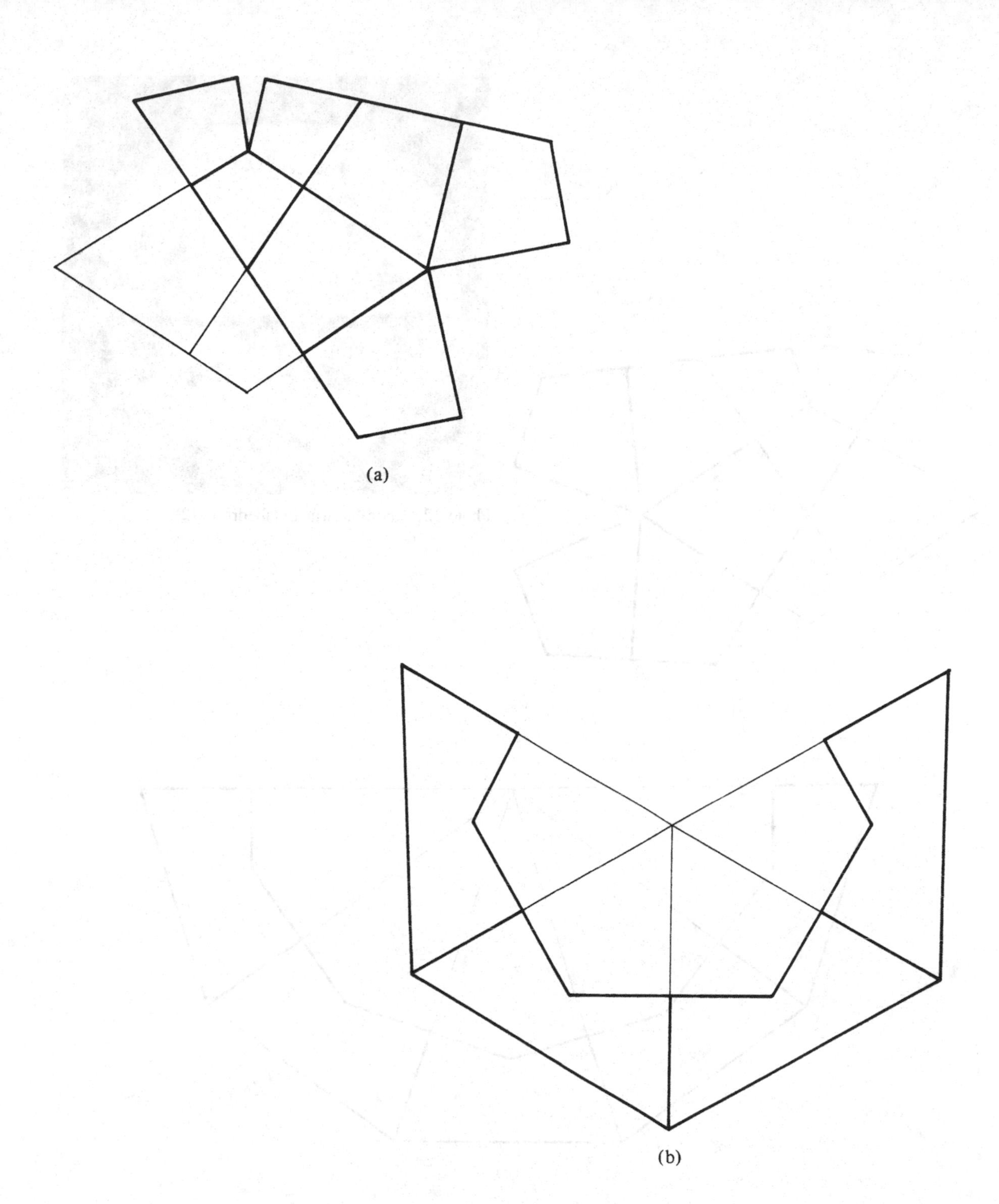

Fig. 15. Patterns for the dual of the cubotahedron (**11**).

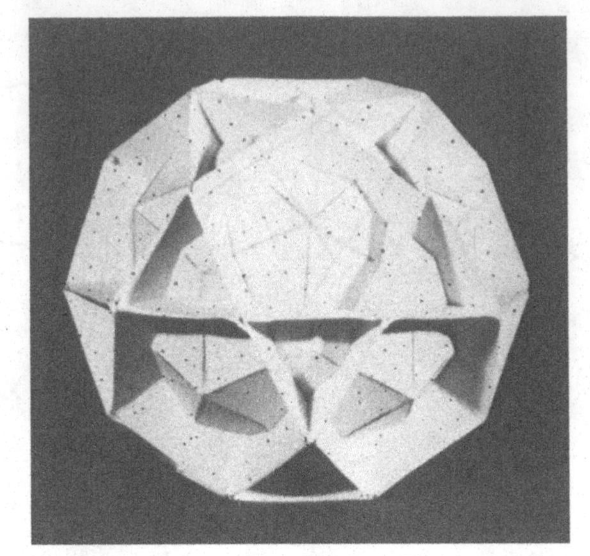

Photo 12. Rhombic triacontahedron (**12**).

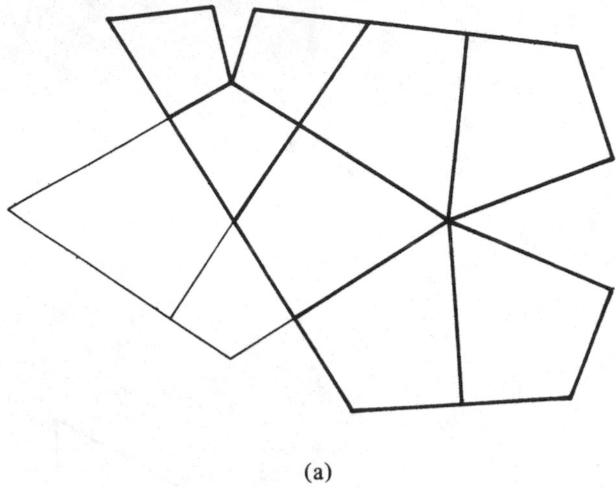

(a)

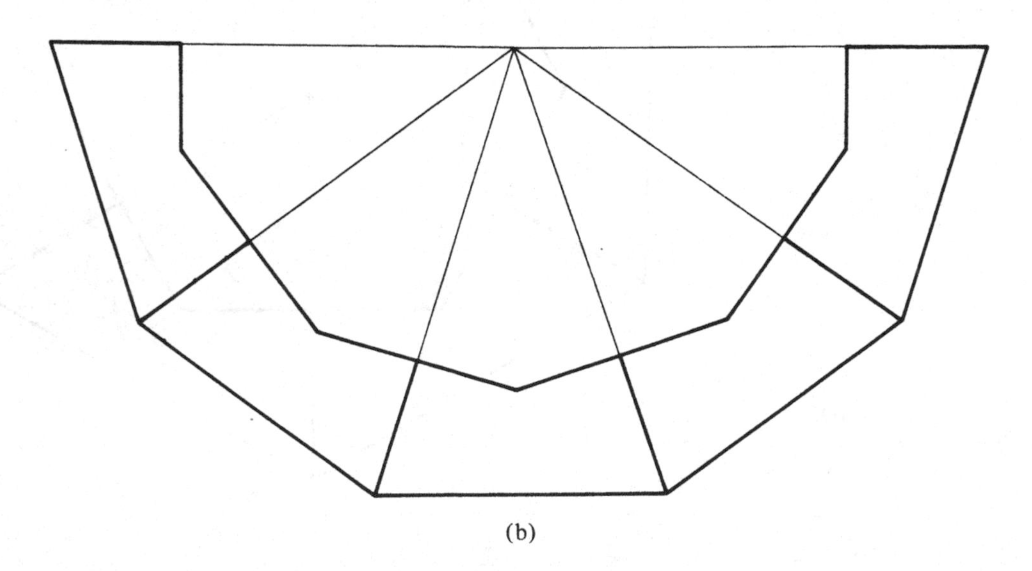

(b)

Fig. 16. Patterns for the dual of the icosidodecahedron (**12**).

Photo 13. Deltoidal icositetrahedron (**13**).

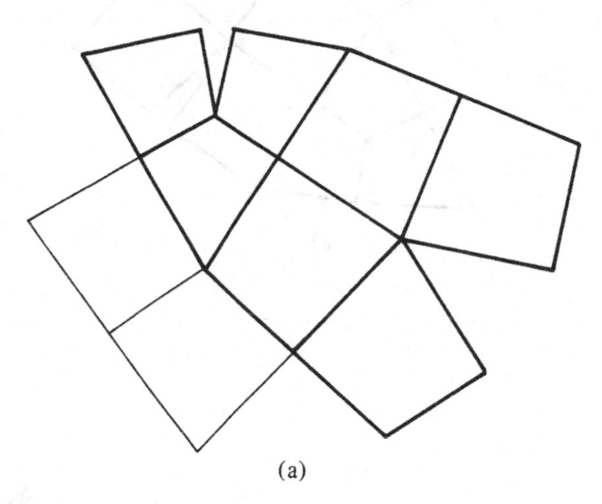

(a)

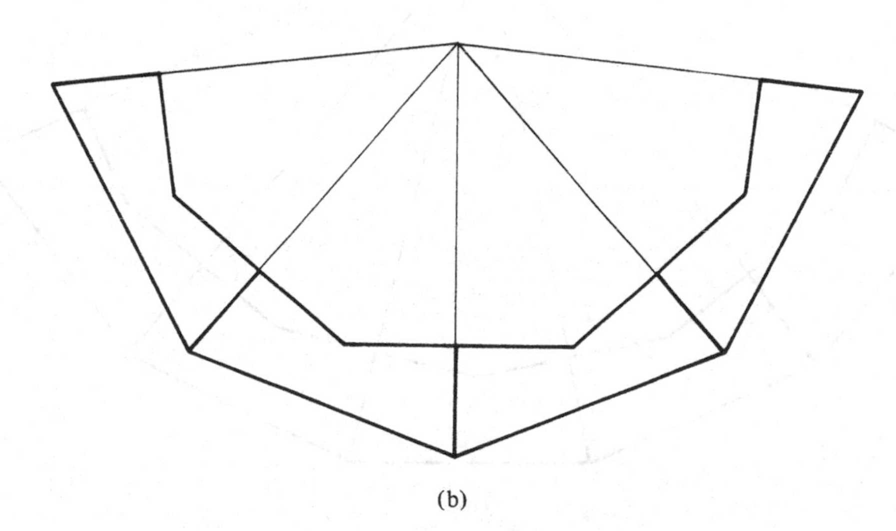

(b)

Fig. 17. Patterns for the dual of the rhombicuboctahedron (**13**).

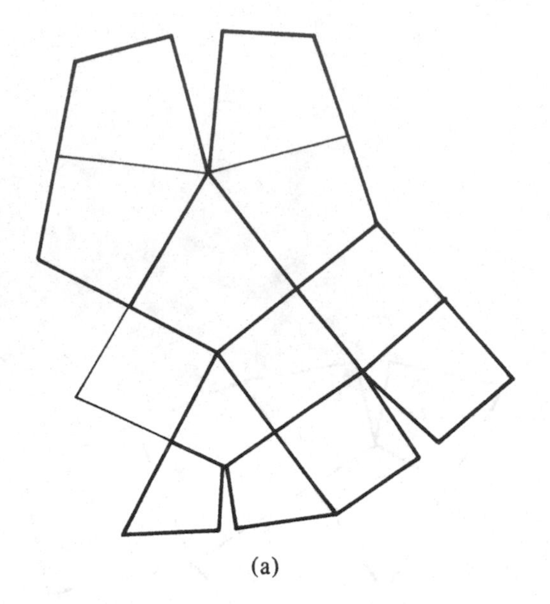

(a)

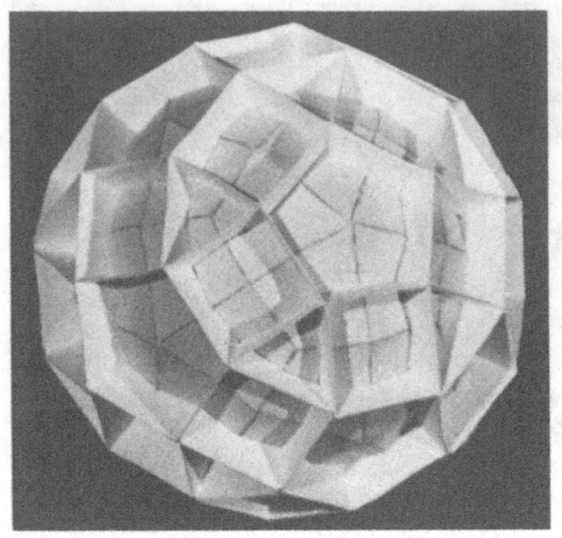

Photo 14. Deltoidal hexecontahedron (**14**).

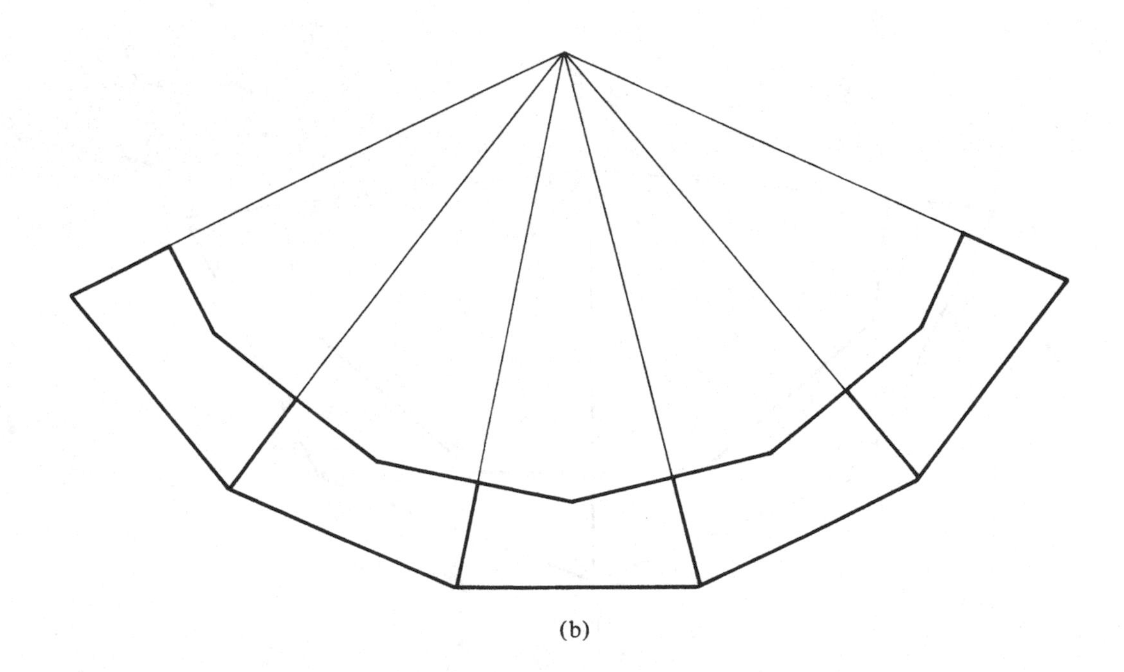

(b)

Fig. 18. Patterns for the dual of the rhombicosidodecahedron (**14**).

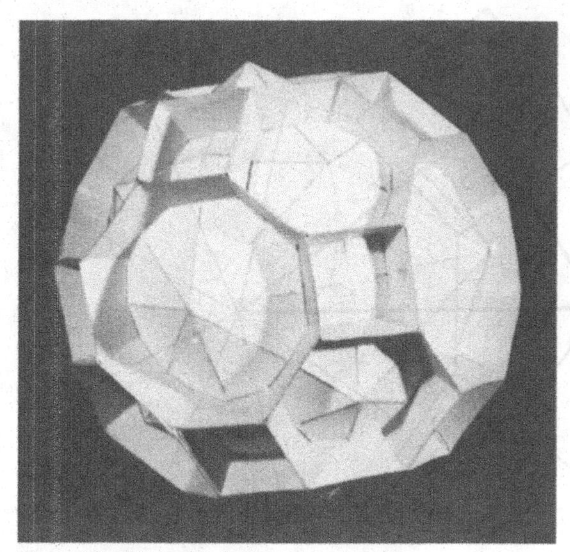

Photo 15. Disdyakisdodecahedron (**15**).

Photo 16. Disdyakistriacontahedron (**16**).

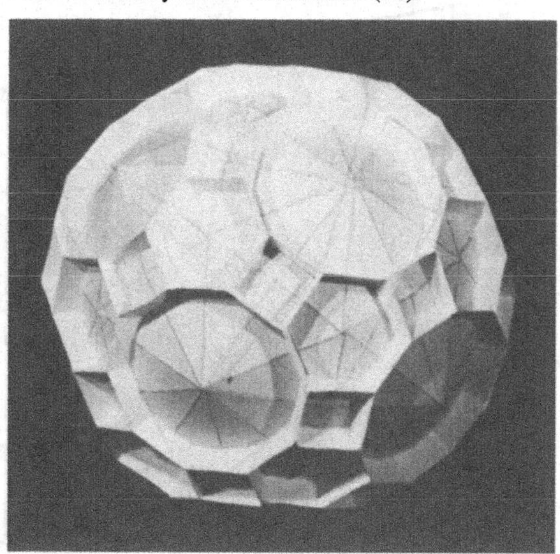

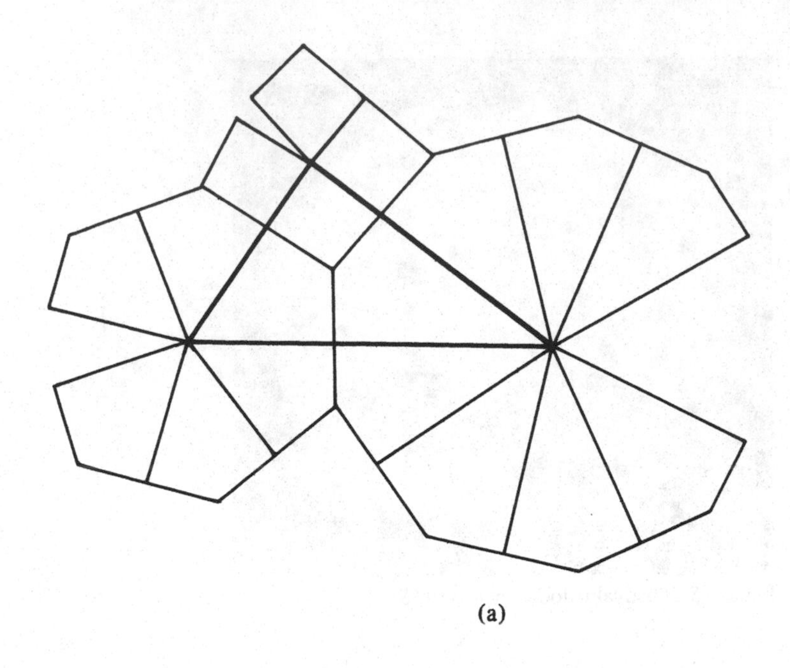
(a)

Fig. 19. Patterns for the dual of the rhombitruncated cuboctahedron (**15**).

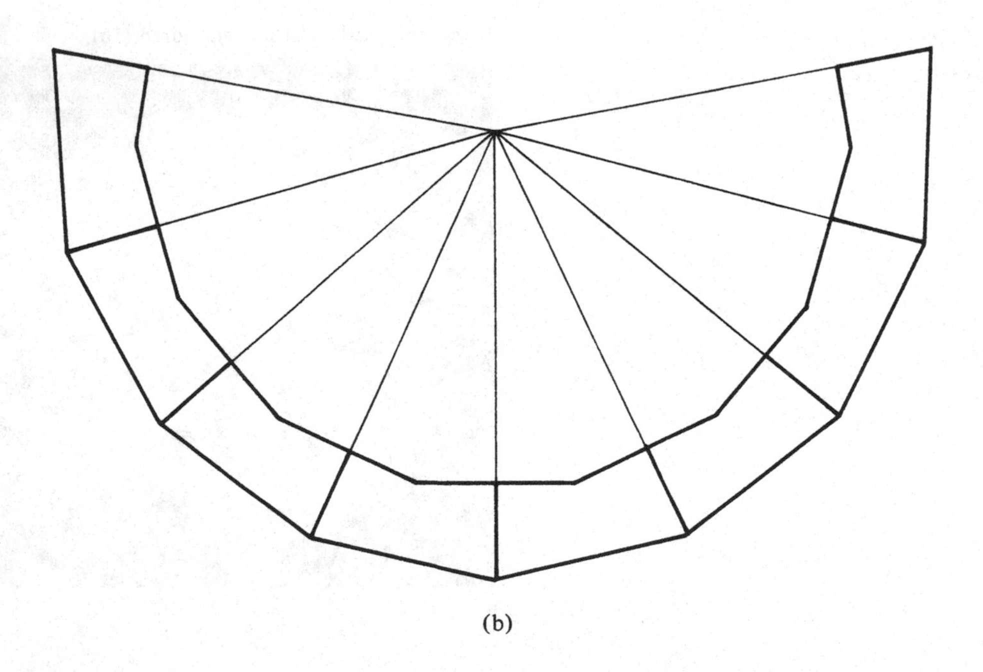
(b)

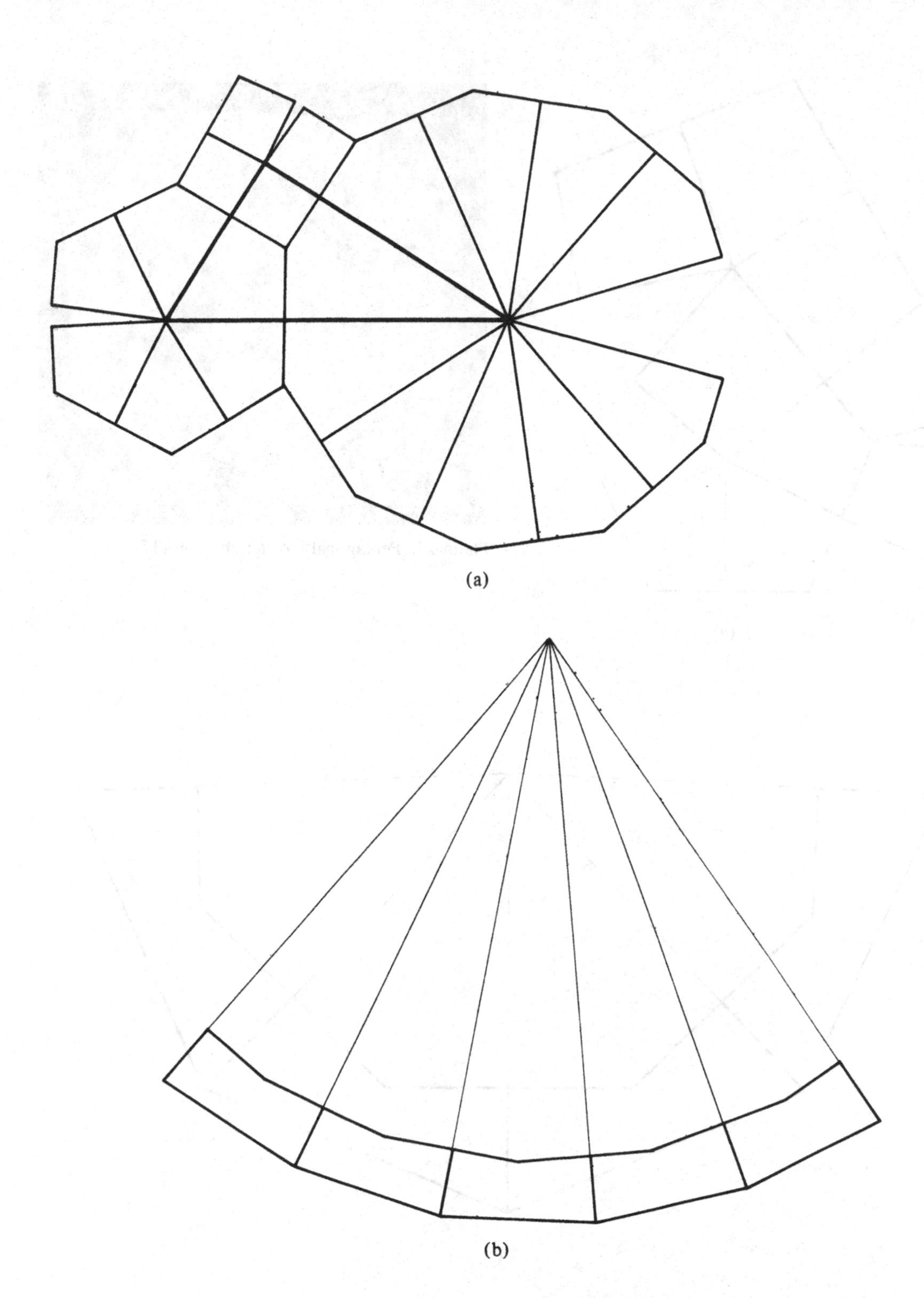

Fig. 20. Patterns for the dual of the rhombitruncated icosidodecahedron (**16**).

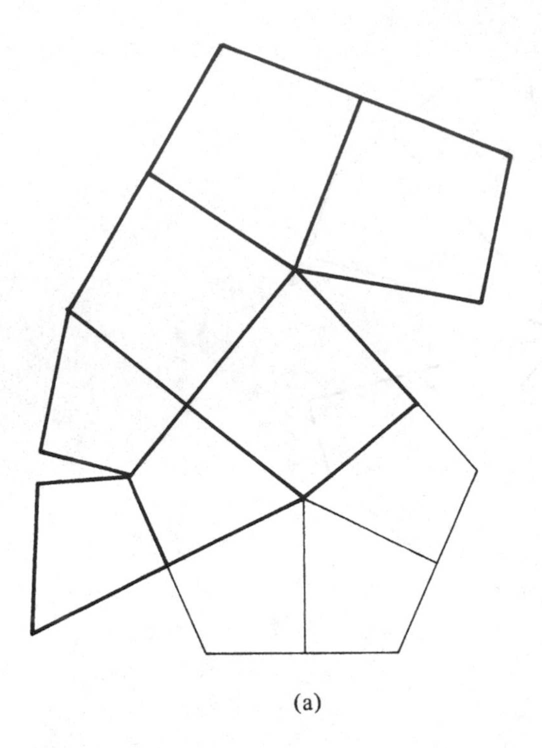

(a)

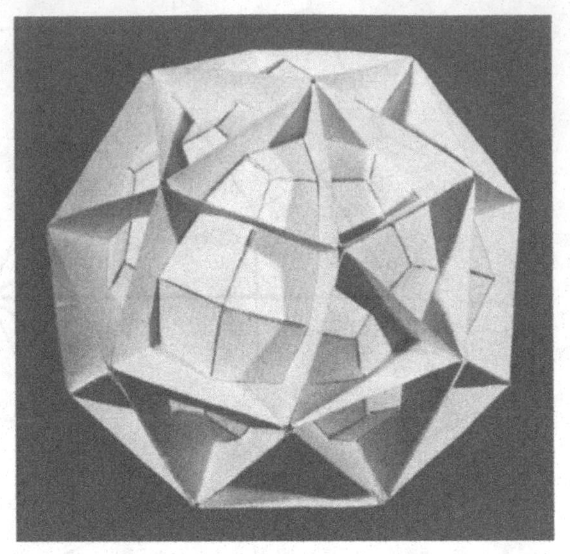

Photo 17. Pentagonal icositetrahedron (**17**).

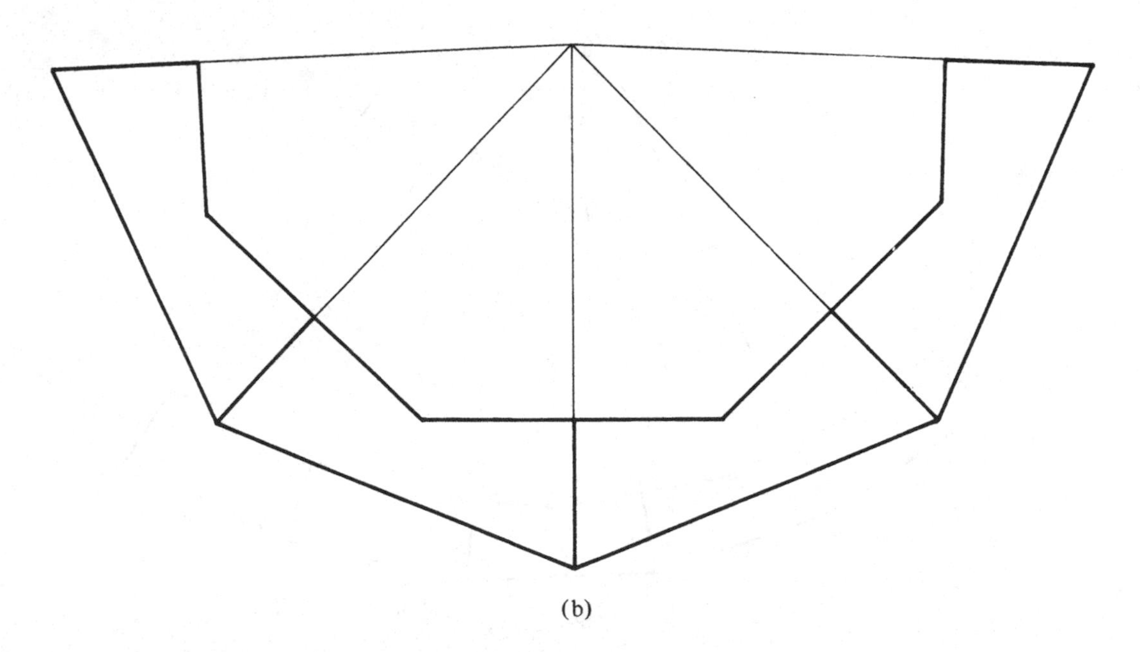

(b)

Fig. 21. Patterns for the dual of the snub cuboctahedron (**17**).

Photo 18. Pentagonal hexecontahedron (**18**).

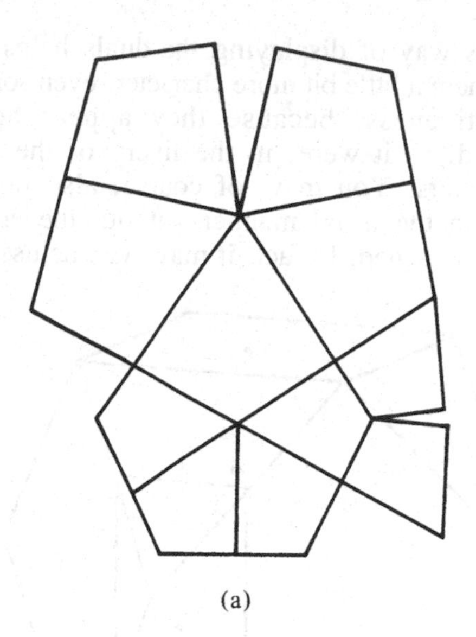

(a)

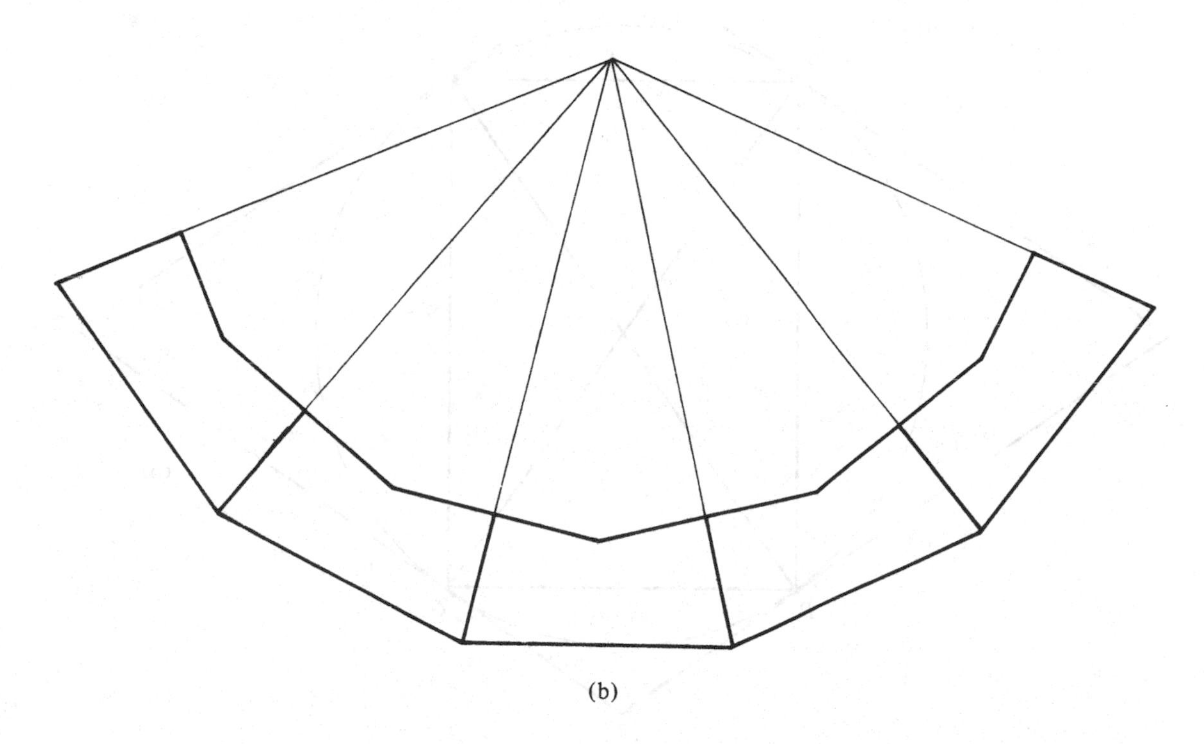

(b)

Fig. 22. Patterns for the dual of the snub icosidodecahedron (**18**).

This way of displaying the duals helps to give them a little bit more character, even some attractiveness, because they appear here dressed, as it were, in the livery of the semiregulars. You may, of course, also make them in the usual manner without the edge model attached. In fact, it may even be useful to make some of them in this way, because as you proceed with the construction you will be seeing aspects not always obvious at first glance. For example, the triakistetrahedron will be seen with a tetrahedron serving as a base solid to which four trigonal pyramids are attached. In making the model, however, you simply omit the inner tetrahedron and cement the pyramids together along their base edges.

The easiest way to find the faces of the convex duals (i.e., the Archimedean or semiregular duals) is to use the so-called Dorman Luke construction. In this construction the faces of convex duals can be derived from the vertex figures of their originals. The vertex figure of a polyhedron is found by connecting the points at a convenient distance, say a unit distance, from any vertex, this distance being measured along edges emanating from that vertex. Figure 23 shows how this applies to the cuboc-

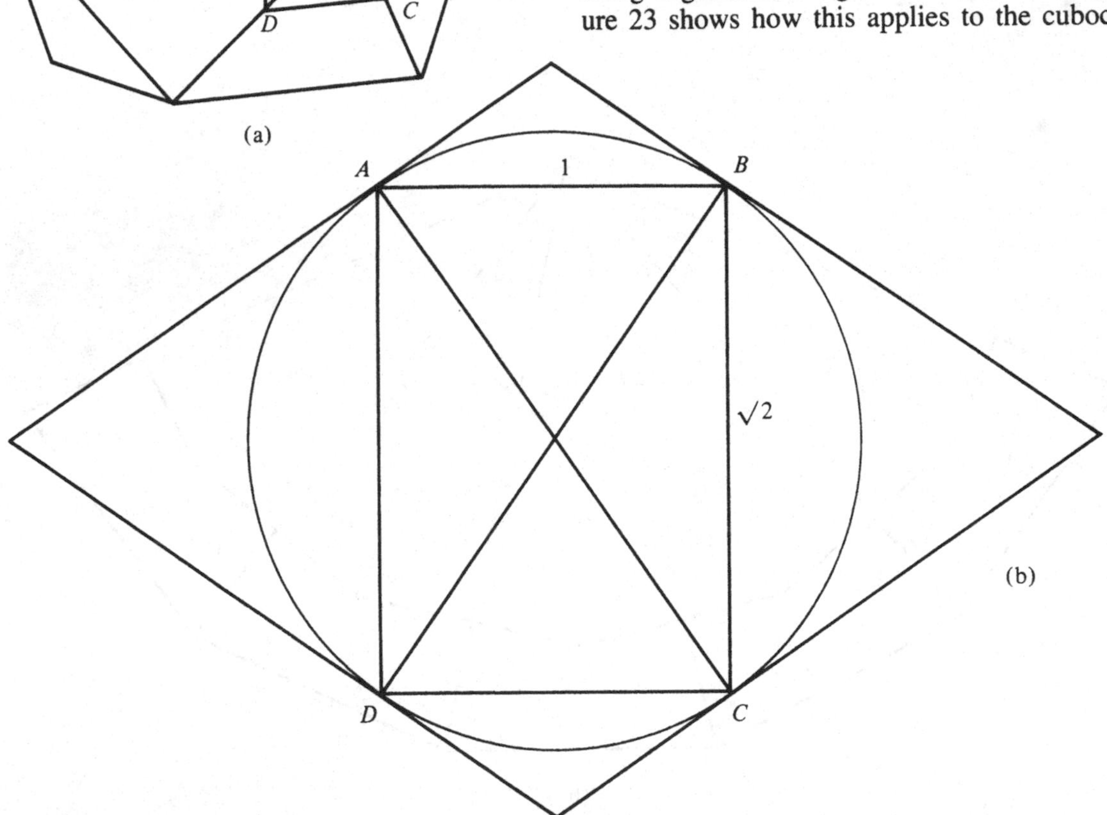

Fig. 23. Dorman Luke construction for the cuboctahedron.

tahedron and its dual, the rhombic dodecahedron. Following the procedure just described (it would be good to have a model of the cuboctahedron in your hands), you will find that the vertex figure is a rectangle whose sides have the measured lengths of one unit and $\sqrt{2}$ units when the edge of the cuboctahedron has a measure of two units. You must now draw this rectangle and then draw the circumcircle around it as shown in Fig. 23. Finally, draw the tangents to this circle at each point where the vertices of the rectangle coincide with the circle. These tangents meet to form a face of the dual. It is an equilateral rhombus, and from the pertinent data the angles of this rhombus can easily be calculated using trigonometric ratios. Because the cuboctahedron has twelve vertices, the rhombic dodecahedron as its dual is appropriately named, having twelve faces, each an equilateral rhombus. For purposes of making a model, the scale of this drawing may be altered to any convenient size, retaining only the angular measures as constant.

Vertex figures were given in *Polyhedron models* under each photo in the book. However, for purposes of calculation, you must have numerical data as well (Table 1). The letter p stands for the kind of polygons surrounding a vertex, and the measures are given in radicals and in decimal form. For purposes of calculation done with an electronic calculator, it is good to have the decimal form to eight significant figures, but for drawing you need at most only three figures or two places of decimals. The edge length assumed here is one unit. Nonconvex values of p are given here for future reference to be used in connection with nonconvex uniform duals.

Figure 24 is an example of how Table 1 is used for the truncated tetrahedron. This has as its vertex figure an isosceles triangle whose sides are each $\sqrt{3} = 1.732$ and whose base is one unit of measure. This triangle is shown with its circumcircle and the tangents to this circle properly drawn. Calculation of the angles for the vertex figure and consequently for the related face of the dual, a face of the triakistetrahedron, can easily be done using the geometrical theorems that apply and some plane trigonometry. The same procedure is operative for the entire set of thirteen semiregular solids. Where the vertex figure is a quadrilateral, as, for example, in the rhombicuboctahedron, you must draw a diagonal to break the vertex figure into two triangles from which the calculation can proceed. For the two convex snub polyhedra, the calculation becomes much more in-

Table 1. *Numerical data for vertex figures*

p convex	Measure in radicals	Measure in decimals	p nonconvex	Measure in radicals	Measure in decimals
3	1	1	$\frac{5}{2}$	$\frac{1}{\tau}$	0.618
4	$\sqrt{2}$	1.414	$\frac{8}{3}$	$(2 - \sqrt{2})^{1/2}$	0.765
5	τ	1.618			
6	$\sqrt{3}$	1.732	$\frac{10}{3}$	$\left(\frac{\sqrt{5}}{\tau}\right)^{1/2}$	1.176
8	$(2 + \sqrt{2})^{1/2}$	1.848			
10	$(\tau\sqrt{5})^{1/2}$	1.902			

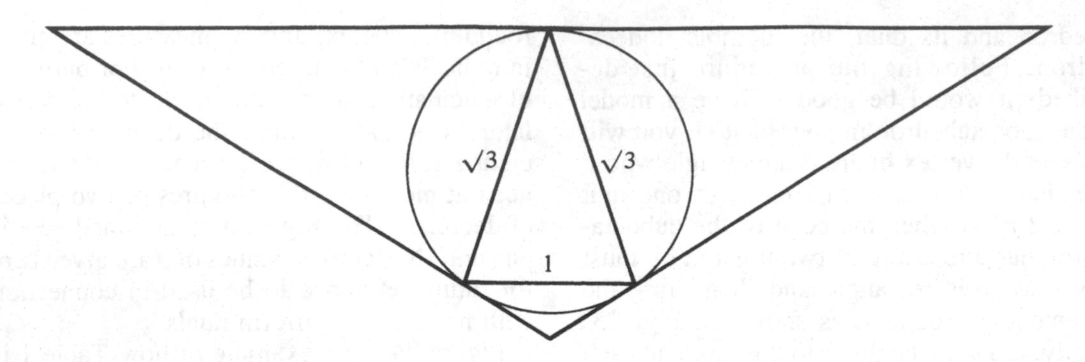

Fig. 24. Dorman Luke construction for the truncated tetrahedron.

volved. It will be left for you as a challenge.

The polar reciprocal formula can also be used to find the face of a semiregular dual. It is a more complicated method to use, but it has important consequences for a deeper undertanding of the convex duals, and even more so for the nonconvex duals that will appear later in this book. Its use will now be illustrated by way of example.

Figure 25d has been drawn in the style of Fig. 13a of *Spherical models* (p. 25) and displays analogous symbolism. This figure is actually a layout of the truncated tetrahedron, from which all the required calculations can easily be done. It would be good if you started keeping your own notebook of numerical data with reference to each polyhedron, because not all the data will be given to you here. Some of the data can be found in *Spherical models* or in other books listed in the references at the end of this book. Much of the work of calculation will be left for you to do by way of mathematical exercises.

First of all, from numerical data you may have recorded in your notebook or have found elsewhere you will know the values to be assigned to ${}_0R$, r_3, a_3, and a_6 in Fig. 25. You can then find ${}_2R_3$, ${}_1R$, and ${}_2R_6$ using the theorem of Pythagoras. Next, by applying the polar reciprocal formula, which in this instance states that $O_3A \cdot {}_2R_3 = {}_0R^2$ and that $O_3B \cdot {}_2R_6 = {}_0R^2$, you can calculate O_3A and O_3B. Also, the angles θ_1 and θ_2 can be found by applying the appropriate trigonometric ratios to respective right-angle triangles. Finally, the lengths a and b can be found using the cosine formula for any triangle, as given in books on plane trigonometry. These are the lengths of the sides of the face triangles of the triakistetrahedron. Once these lengths are known, this face triangle can be drawn, and the angles of this triangle can be calculated using the sine and cosine formulas from plane trigonometry. The scale of this face can then be altered to any convenient length, retaining only the angles constant.

Once you understand the fact that Fig. 25 is a layout, you can make analogous drawings for any or all of the other semiregular solids and from these drawings calculate the lengths of sides and hence the measures of the face angles of the duals. If faces of a dual happen to be quadrangular in shape, you must use a diagonal to form the triangles that compose this quadrangle and then proceed to make the required layout for such a situation. Figure 26 shows the layout for the cuboctahedron and how its dual, the rhombic dodecahedron, is related to it as an example of such a situation. You notice that this is simpler than the triakistetrahedron, but the latter is actually more typical.

Admittedly, the polar reciprocal construction method for finding the face of a dual is

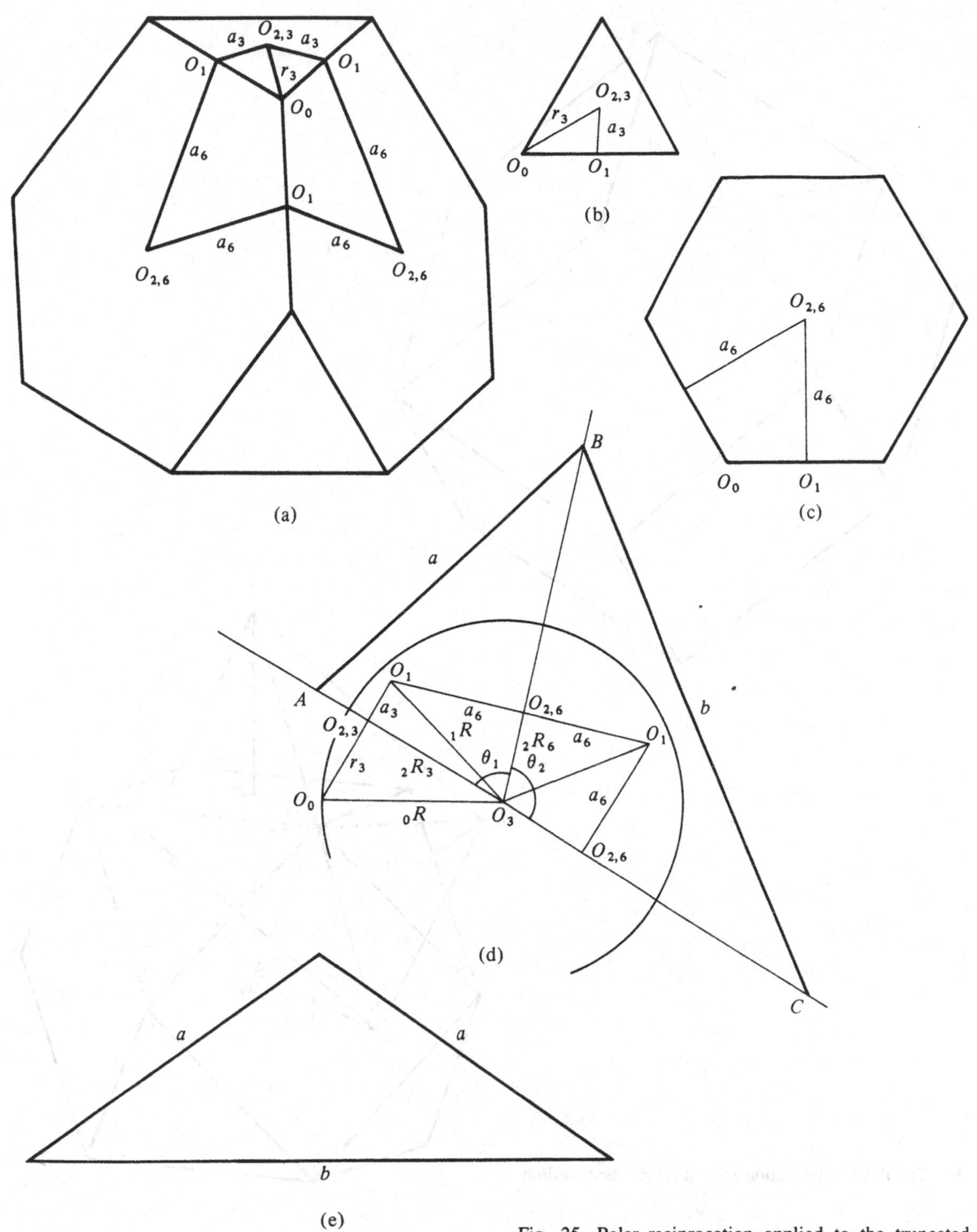

Fig. 25. Polar reciprocation applied to the truncated tetrahedron.

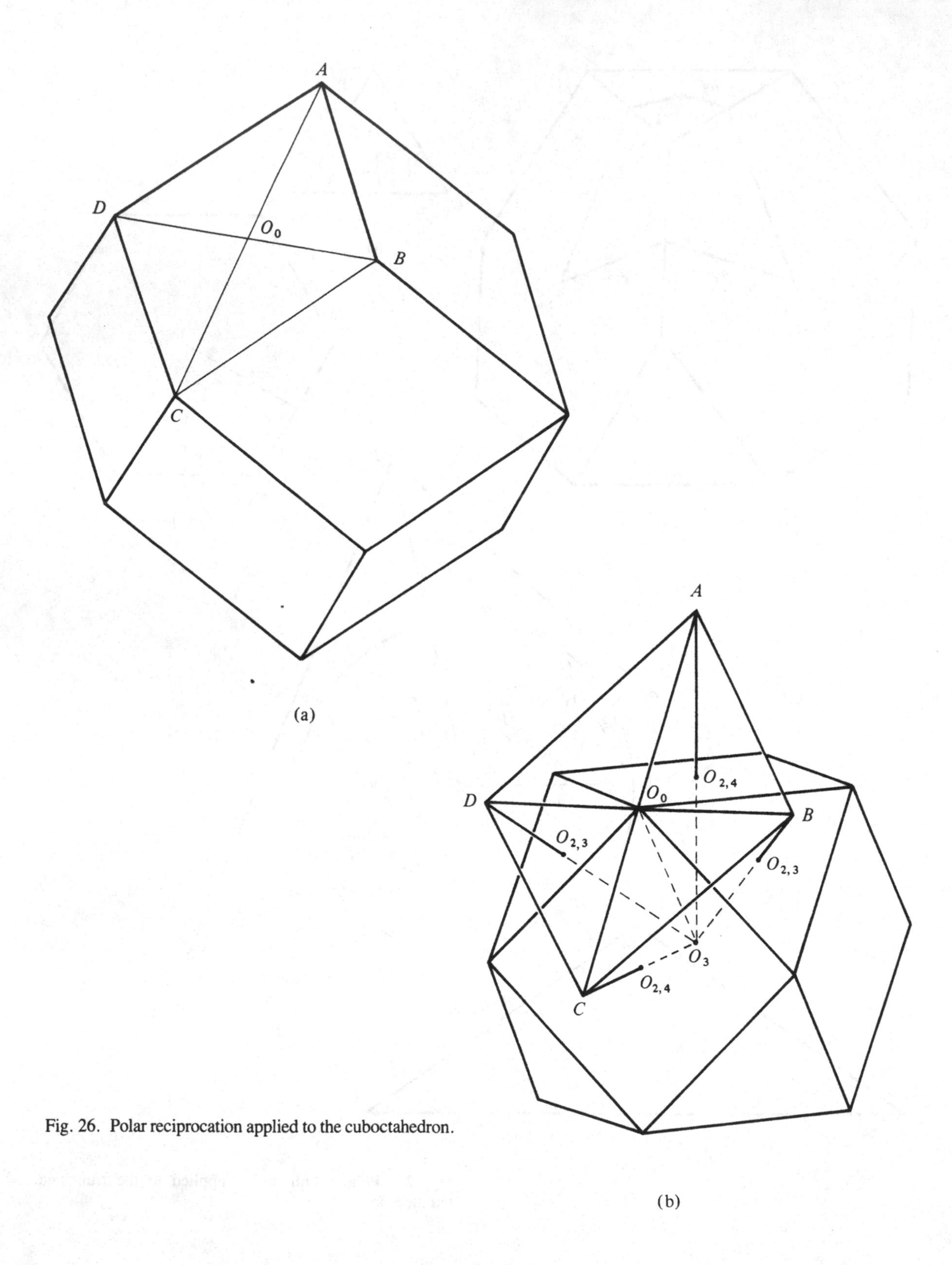

Fig. 26. Polar reciprocation applied to the cuboctahedron.

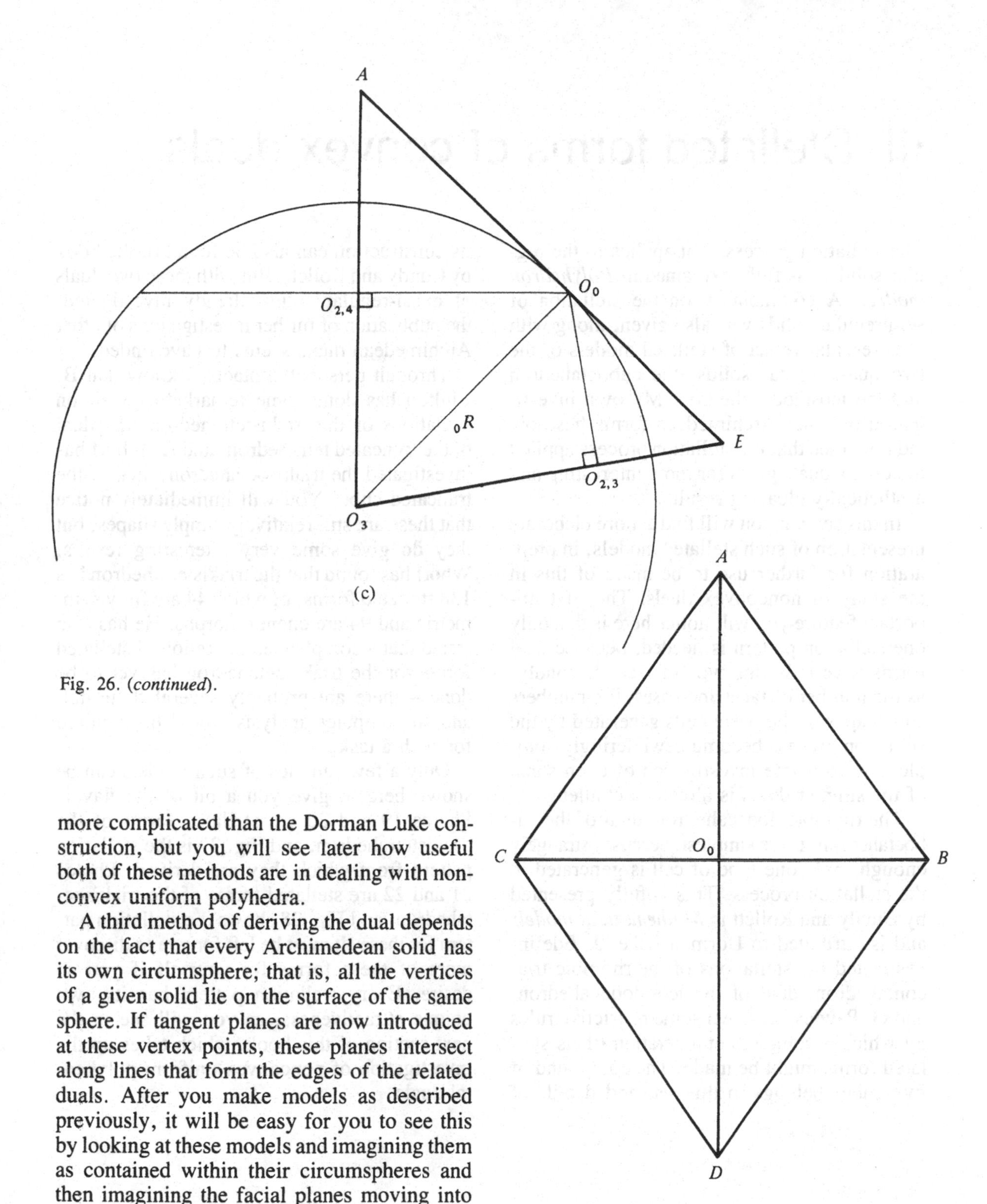

Fig. 26. (*continued*).

more complicated than the Dorman Luke construction, but you will see later how useful both of these methods are in dealing with nonconvex uniform polyhedra.

A third method of deriving the dual depends on the fact that every Archimedean solid has its own circumsphere; that is, all the vertices of a given solid lie on the surface of the same sphere. If tangent planes are now introduced at these vertex points, these planes intersect along lines that form the edges of the related duals. After you make models as described previously, it will be easy for you to see this by looking at these models and imagining them as contained within their circumspheres and then imagining the facial planes moving into the respective tangent planes.

III. Stellated forms of convex duals

The stellation process as it applies to the regular solids was fully explained in *Polyhedron models*. A commentary on the stellation of semiregular solids was also given, along with a representative set of stellated models of the two quasi-regular solids, the cuboctahedron and the icosidodecahedron. My own investigation of other Archimedean forms has now led me to see that the stellation process applied to convex duals gives far more interesting and aesthetically pleasing results.

In this section you will find a more elaborate presentation of such stellated models, in preparation for further use to be made of this in the study of nonconvex duals. The first important feature you will notice here is that only one stellation pattern is needed, because dual forms have only one type of face. Secondly, as the number of faces increases, the numbers and shapes of the basic cells generated by the stellation process become bewilderingly complex. A complete investigation of even some of the simpler duals is already a challenge.

The rhombic dodecahedron, dual of the cuboctahedron, is the simplest, because, strangely enough, only one type of cell is generated in the stellation process. This is fully presented by Cundy and Rollett in *Mathematical models* and is attributed to Dorman Luke. J. Ede investigated the stellations of the rhombic triacontahedron, dual of the icosidodecahedron, and G. Pawley set down some restrictive rules by which a complete enumeration of its stellated forms might be made. The compound of five cubes belongs to this set, and details of its construction can also be found in the book by Cundy and Rollett. But with these two duals of quasi-regular solids already investigated, the publication of further investigations of other Archimedean duals seems to have ended.

Through personal contacts, I know that B. Chilton has done some remarkable work on stellations of the triakistetrahedron, the dual of the truncated tetrahedron, and R. Whorf has investigated the triakisoctahedron, dual of the truncated cube. You will immediately notice that these are still relatively simple shapes, but they do give some very interesting results. Whorf has found that the triakistetrahedron has 138 stellated forms, of which 44 are fully symmetric and 94 are enantiomorphs. He has also stated that a complete enumeration of stellated forms for the triakisoctahedron has yet to be done – there are probably several thousand, and so computer analysis would be required for such a task.

Only a few samples of such models can be shown here to give you a bit of the flavor. Photos 19 and 20 are stellated forms of the triakistetrahedron, and Fig. 27 is the stellation pattern from which they are derived. Photos 21 and 22 are stellated forms of the triakisoctahedron, and Fig. 28 shows the stellation pattern for these. It must be left for you to discover more of these forms for yourself if you so desire. Every stellation pattern has its own unique attractiveness, as you will see in the next section of this book, which takes up the investigation of nonconvex uniform polyhedral duals.

Photo 19. First stellation of the triakistetrahedron.

Photo 20. Final stellation of the triakistetrahedron.

Fig. 27. Stellation pattern for the triakistetrahedron.

Photo 21. A stellation of the triakisoctahedron.

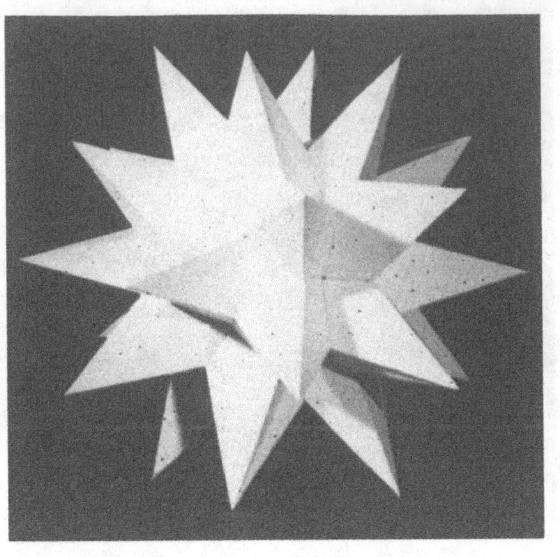

Photo 22. Another stellation of the triakisoctahedron.

Fig. 28. Stellation pattern for the triakisoctahedron.

IV. The duals of nonconvex uniform polyhedra

As mentioned previously, my own investigations of stellated forms of convex or Archimedean duals finally led me to see the close relationship that exists between the stellation process and duality of nonconvex forms. The most notable examples here belong to the set of polyhedra known as the Kepler–Poinsot solids. These will provide you with a good introduction to this topic.

The regular nonconvex uniform polyhedra and their duals

The four regular nonconvex uniform polyhedra, also called the Kepler–Poinsot solids, were shown in *Polyhedron models* as **20**, **21**, **22**, and **41**. Because they belong to the complete set of nonconvex uniform duals, they are shown again here in Photos 23, 24, 25, and 26. To avoid confusion, all polyhedron models will henceforth be referred to by their boldface numbers as used in *Polyhedron models*. A closer study of the Kepler–Poinsot solids reveals that each has its dual within the same set of four, just as the five regular convex solids have their duals within the same set of five. Figure 29 is the stellation pattern for the dodecahedron, and Fig. 30 is that for the icosahedron. Figures 31 through 34 show the Dorman Luke construction applied to respective vertex figures and a facial plane for each. You see that this facial plane is embedded in respective stellation patterns. The shaded portions show the parts of each facial plane that will be visible on the outside of the solid. With nonconvex solids, you see that facial planes intersect before they meet again along edges of the solid, two faces sharing an edge in common. In the same way, the edges also intersect each other before they meet again at each vertex. Does it seem strange to you that the dual

Photo 23. Great dodecahedron (**20**).

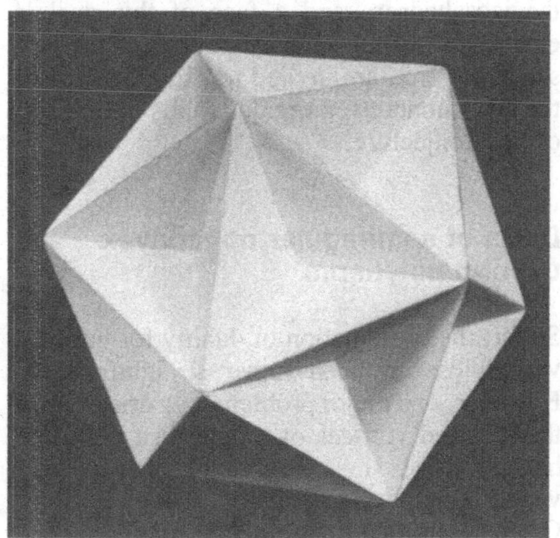

Photo 24. Small stellated dodecahedron (**21**).

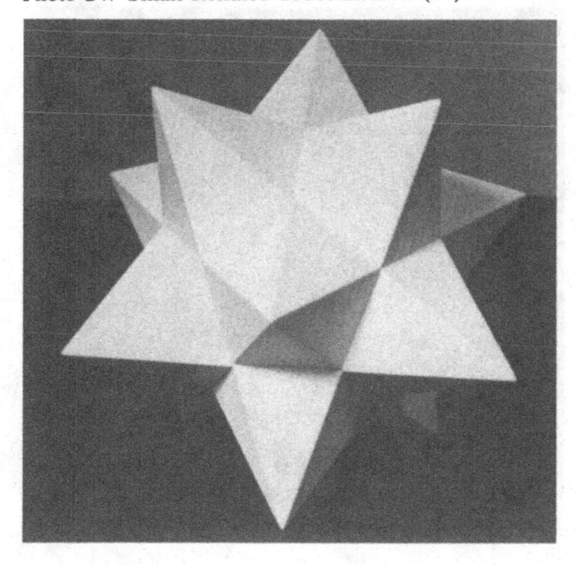

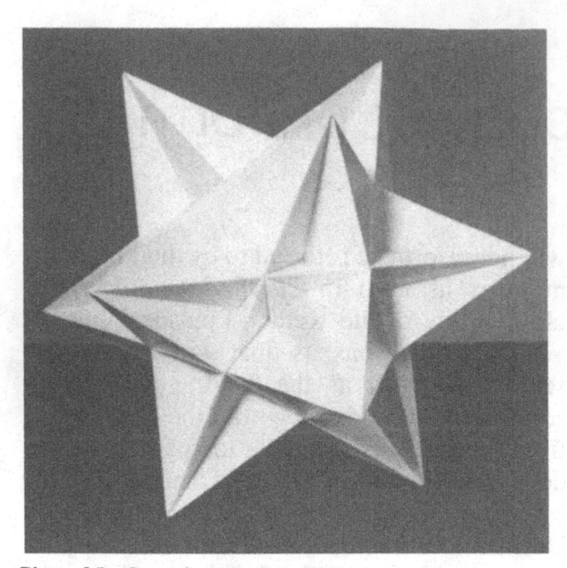

Photo 25. Great icosahedron (**22**).

Photo 26. Great stellated dodecahedron (**41**).

of the great stellated dodecahedron turns out to be a stellated form of the icosahedron? This strangeness disappears if you look first of all at the number and kinds of faces and vertices involved here. The great stellated dodecahedron has twelve star faces, called also pentagrammic faces, and so its dual must have twelve pentagrammic vertices. These are exactly the number and kinds of vertices that belong to the great icosahedron. The other regular nonconvex solids remain in the stellations of the dodecahedron, because pentagon and pentagram shapes have the same number of sides, namely five. The clearest indication of the interchange of the kinds of faces and vertices is shown in the so-called Schläfli symbol. These are, for **20**: (5⁄2, 5); for **21**: (5, 5⁄2); for **22**: (5⁄2, 3); for **41**: (3, 5⁄2). The first numeral names the kind of face, and the second names the kind of vertex for each. Thus, **20** and **21** are duals of each other, and also **22** and **41** are duals of each other.

The relationship enunciated in the Introduction as a conjecture is beautifully illustrated by these four nonconvex uniform polyhedra. For example, the convex hull of the great stellated dodecahedron is a regular dodecahedron. The dual of the dodecahedron is the icosahedron. Hence, the dual of the great stellated dodecahedron is a stellated form of the icosahedron, and a face of this dual is embedded in the stellation pattern of the icosahedron. You are invited to examine each of the other nonconvex regular solids in the light of this conjecture.

Duals of semiregular nonconvex uniform polyhedra

So far, the presentation of duality for nonconvex solids has been rather straightforward. However, one major problem now arises when the required vertices of a dual turn out to be hidden vertices (i.e., vertices that are no longer visible from the outside of a model). Both Cundy and Rollett in *Mathematical models* and

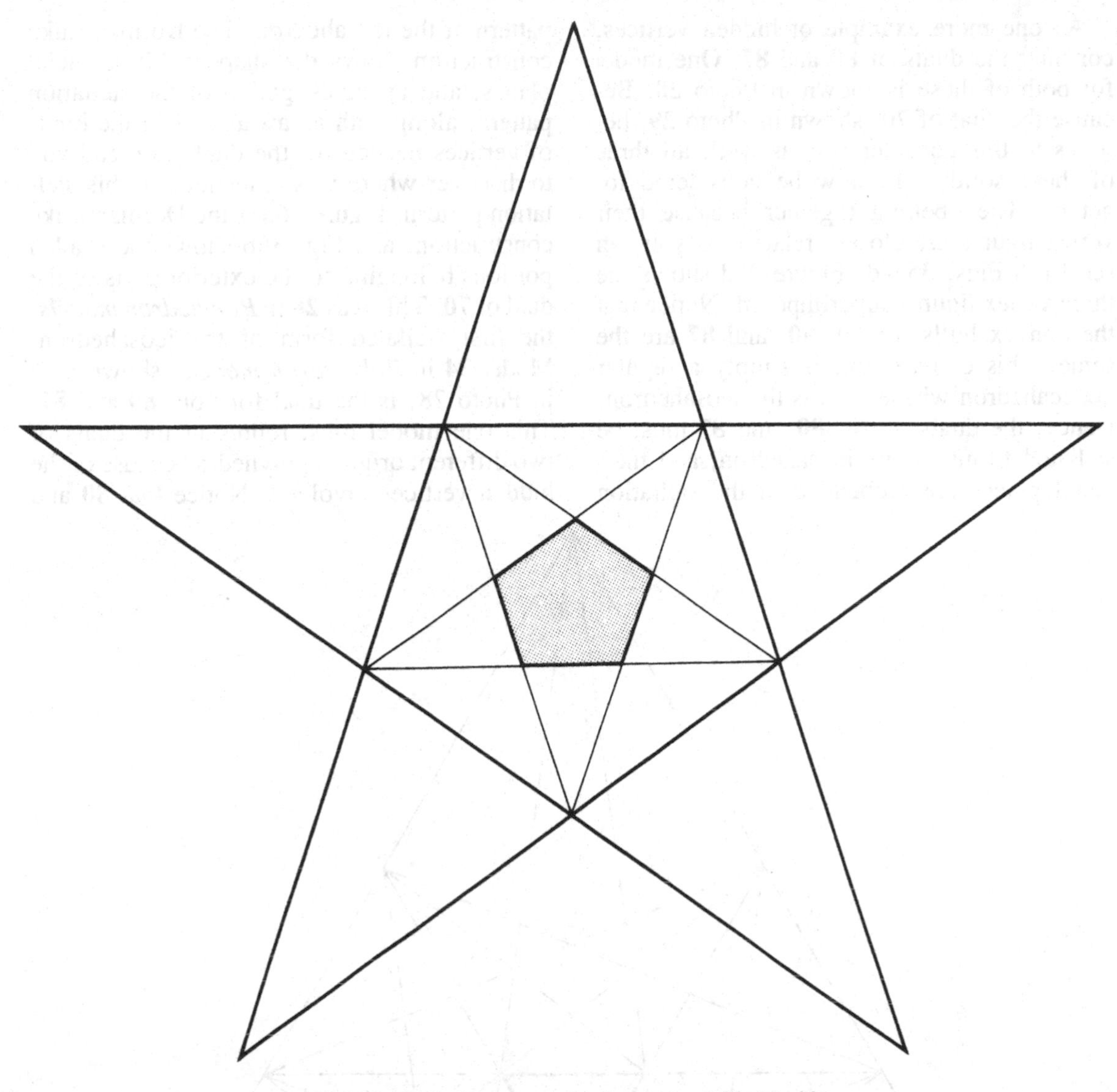

Fig. 29. Stellation pattern for the duals of **20**, **21**, **22**.

Coxeter in *Regular polytopes* pointed out this phenomenon with respect to the dual of **73**. This uniform polyhedron has a convex hull that is an icosidodecahedron, whose dual is the rhombic triacontahedron; so the dual of **73** is a stellated form of the rhombic triacontahedron. But the pentagrammic vertices lie hidden directly under the pentagonal vertices. Photo 27 shows the model, in which only the pentagonal vertices are visible. It is important to notice that the pentagrammic faces of **73** are farther out from the center of symmetry of the solid than are the pentagon faces. If the polar reciprocal formula were to be applied, you would obtain the exact positions of these vertices for the dual relative to the incenters of respective faces of the original. This same phenomenon of hidden vertices will be coming up over and over again in the investigation of other duals of nonconvex uniform polyhedra.

As one more example of hidden vertices, consider the duals of **80** and **87**. One model for both of these is shown in Photo 28. Because the dual of **70**, shown in Photo 29, belongs to this consideration as well, all three of these solids will now be considered together. They belong together because their vertex figures are closely related, as you can see from Figs. 35a–d. Figure 35d shows the three vertex figures superimposed. Notice that the convex hulls for **70**, **80**, and **87** are the same. This convex hull is simply a regular dodecahedron whose dual is the icosahedron. Hence, the duals of **70**, **80**, and **87** must be stellated forms of the icosahedron, and their facial planes are embedded in the stellation pattern of the icosahedron. The Dorman Luke construction shows the shape of these facial planes, and an investigation of the stellation pattern, along with an awareness of the kinds of vertices needed for the dual, can lead you to discover where it is embedded in this stellation pattern. Figure 36a is the Dorman Luke construction, and Fig. 36b shows the shaded portions belonging to the exterior parts of the dual of **70**. This was **26** in *Polyhedron models*, the first stellated form of the icosahedron. Model **34** in *Polyhedron models*, shown here in Photo 28, is the dual for both **80** and **87**. This one model must represent the duals of two different original polyhedra because of the hidden vertices involved. Notice that **80** and

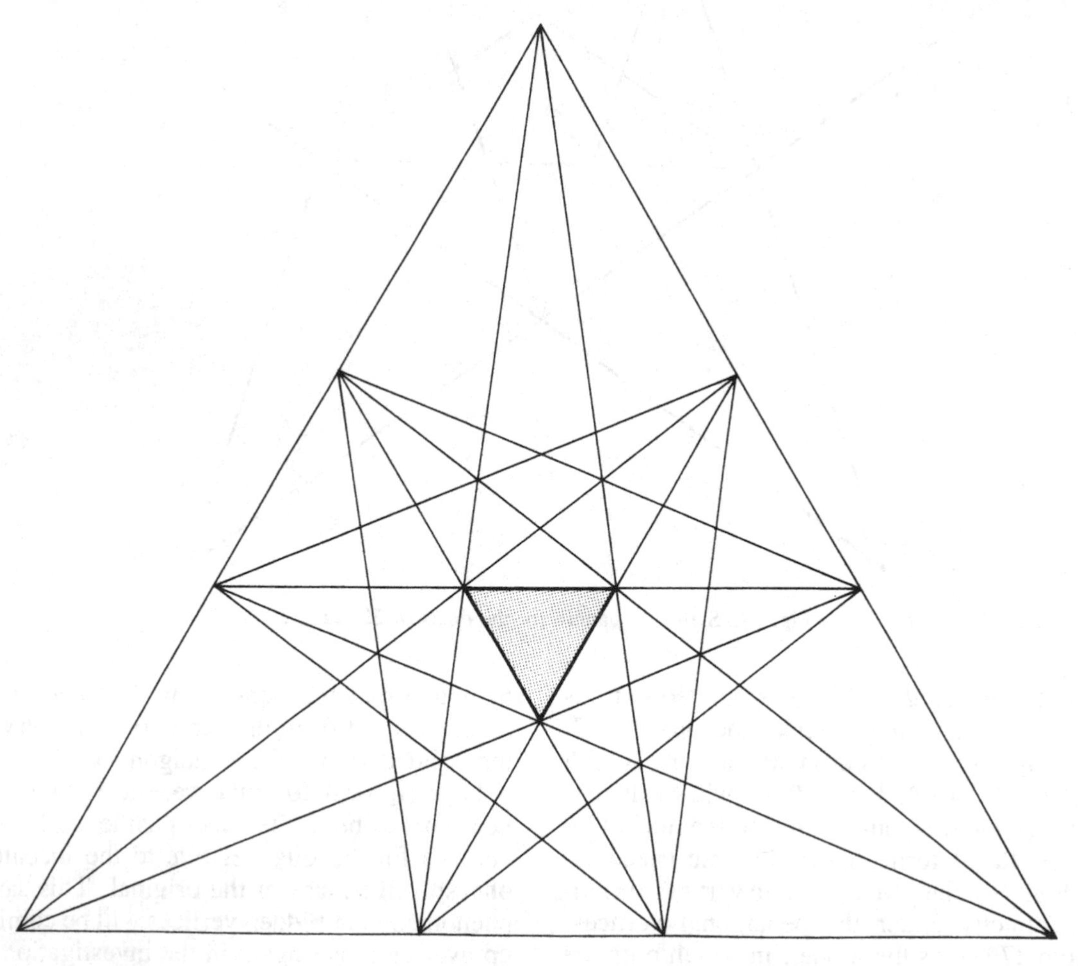

Fig. 30. Stellation pattern for duals of **41**, **70**, **80**, **87**.

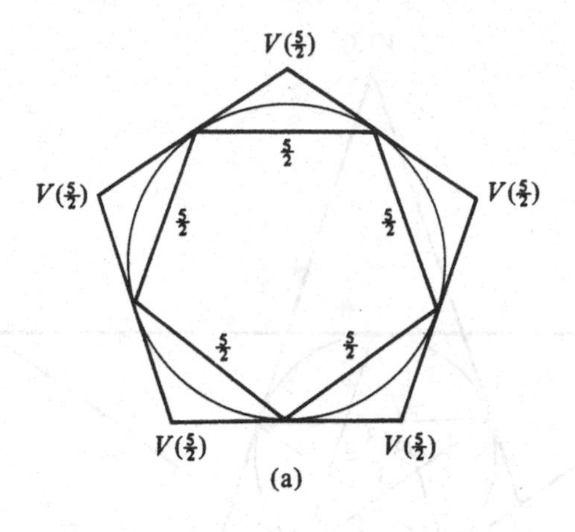

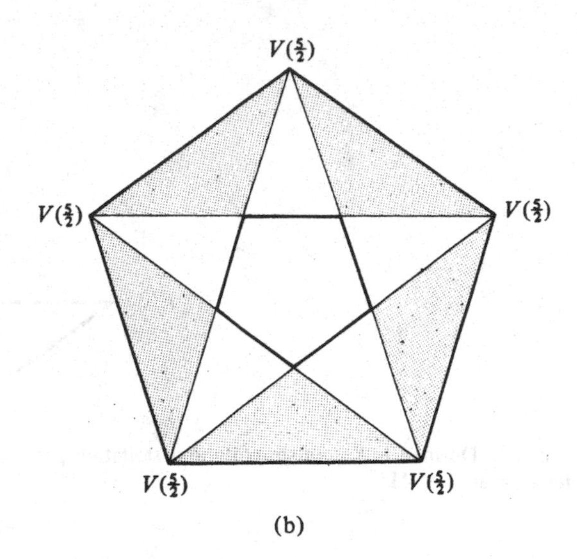

Fig. 31. Dorman Luke construction and stellation pattern for dual of **20**.

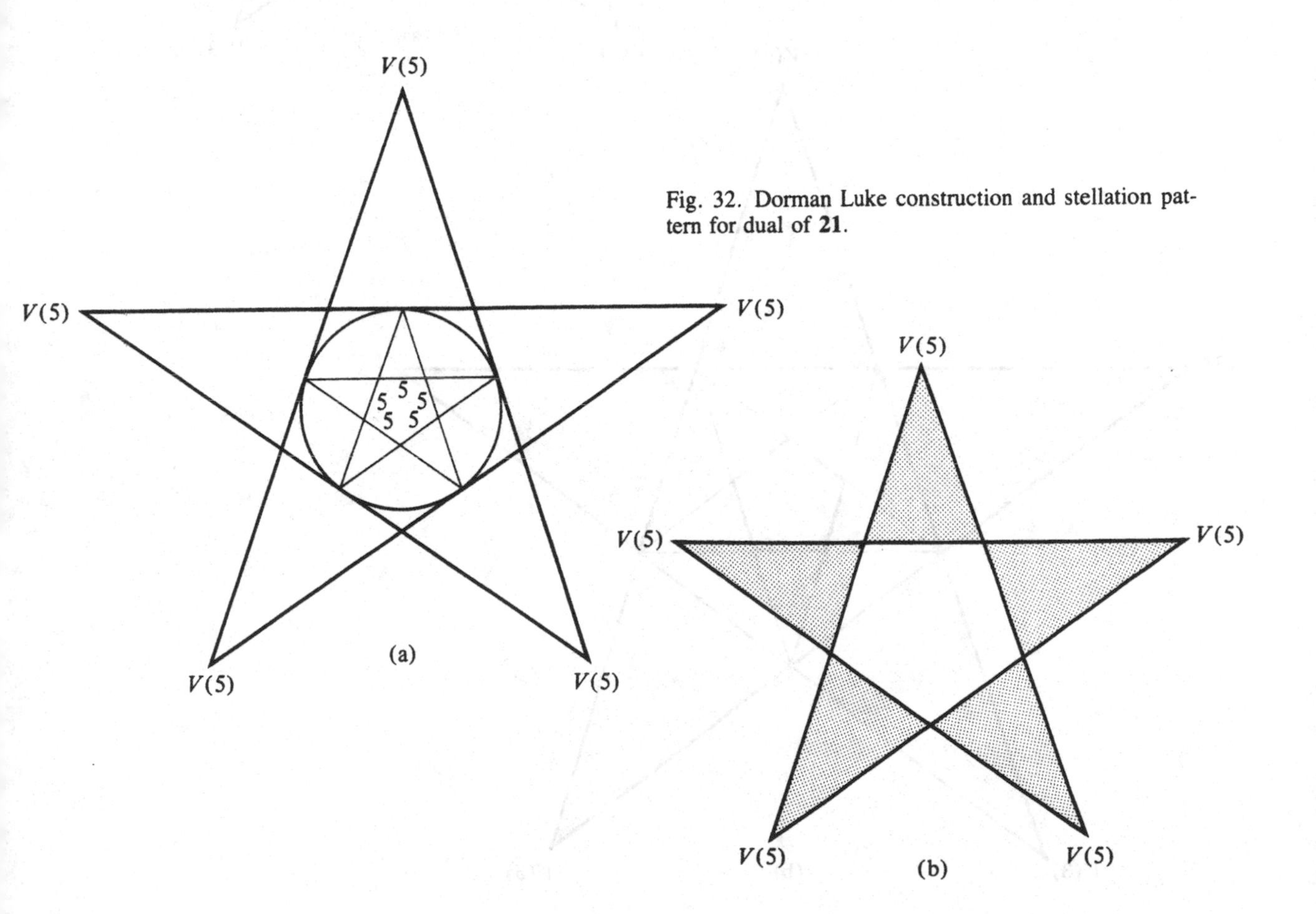

Fig. 32. Dorman Luke construction and stellation pattern for dual of **21**.

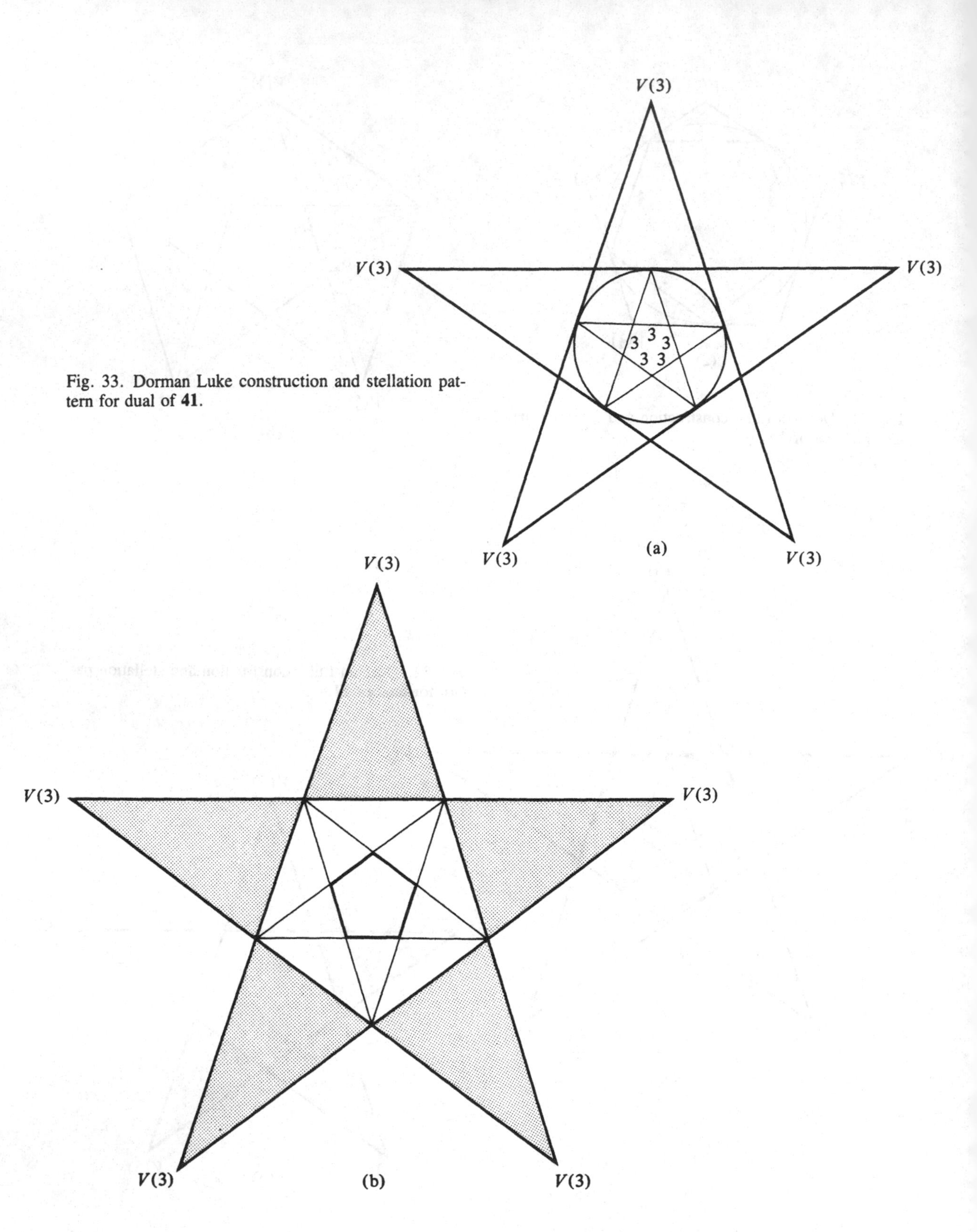

Fig. 33. Dorman Luke construction and stellation pattern for dual of **41**.

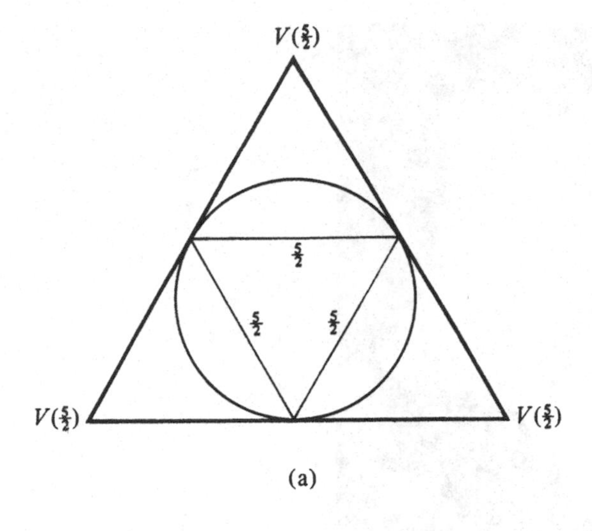

87 have the same pentagonal faces, and hence their duals have the same pentagonal vertices. But **80** has pentagrammic faces, whereas **87** has triangle faces, and hence their duals must have pentagrammic vertices for the one and trigonal vertices for the other. The Dorman Luke constructions shown in Figs. 37a and 38a reveal the shapes of the required faces of these duals and the positions of the vertices on these faces. When this information is transferred to the stellation pattern of the icosahedron and then applied to a model, you will find the pentagrammic vertices hidden in one part of the interior of the model and the trigonal vertices in another part, but both types are covered by the intersecting planes of the pentagonal spikes of the dual.

Fig. 34. Dorman Luke construction and stellation pattern for dual of **22**.

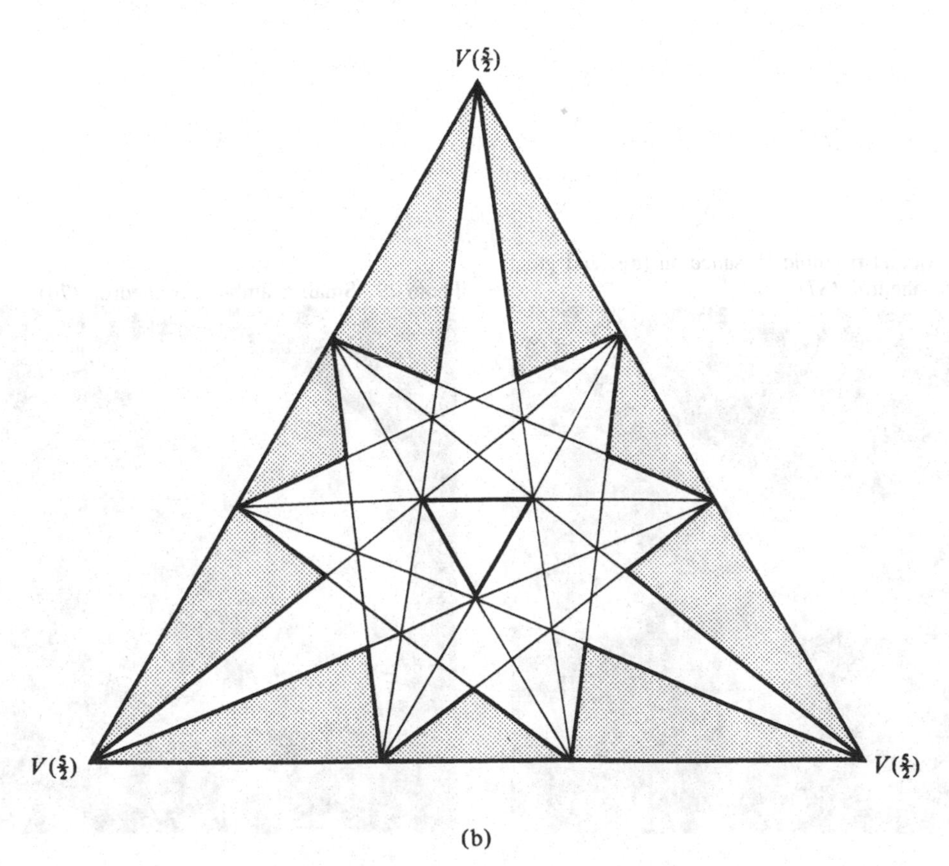

Photo 27. Medial rhombic triacontahedron (**73**).

Photo 28. Medial triambic icosahedron (**80**) and great triambic icosahedron (**87**).

Photo 29. Small triambic icosahedron (**70**).

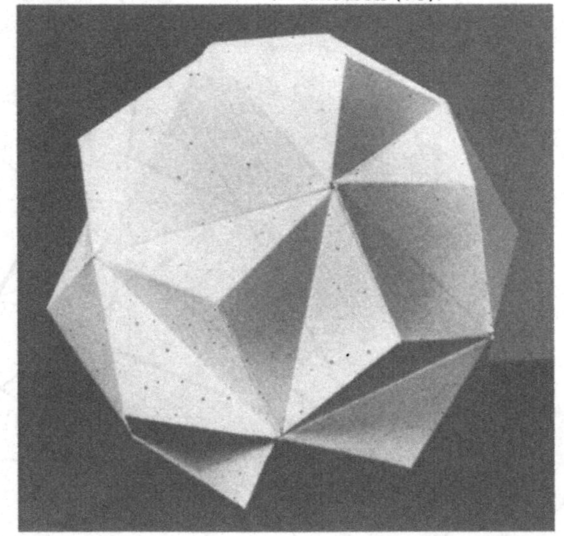

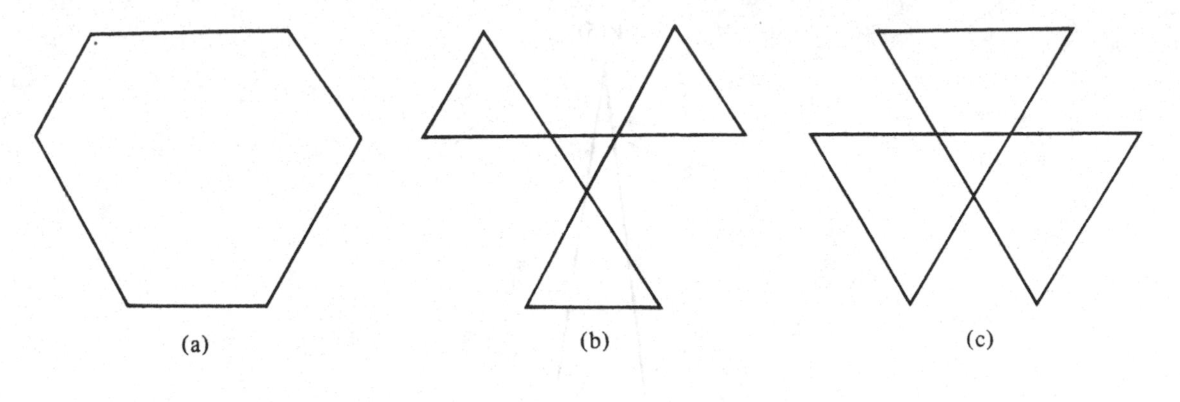

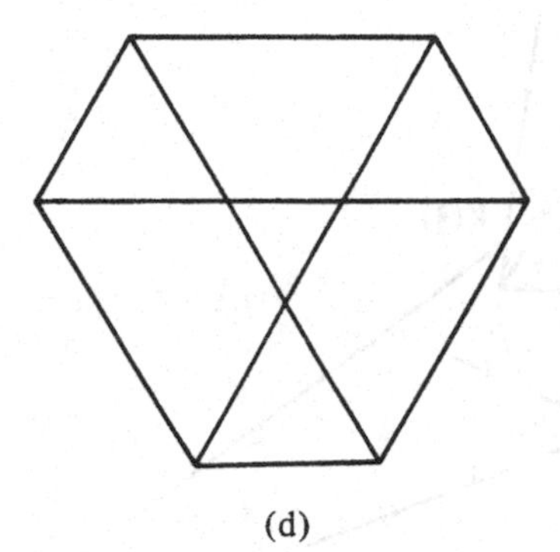

Fig. 35. Vertex figures for **70**, **80**, and **87**, all superimposed.

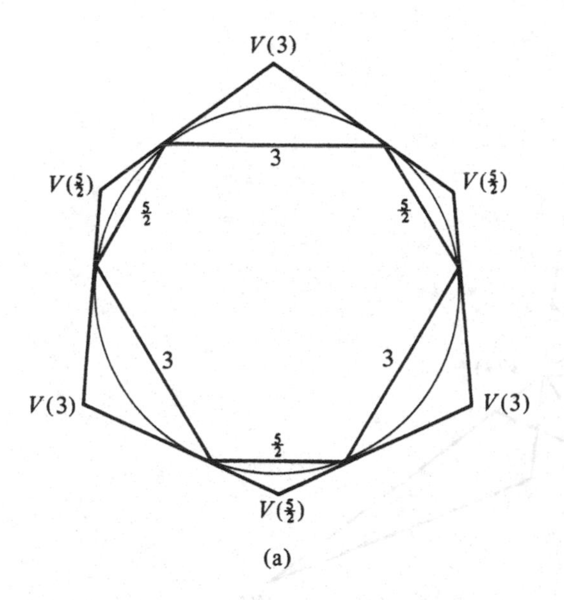

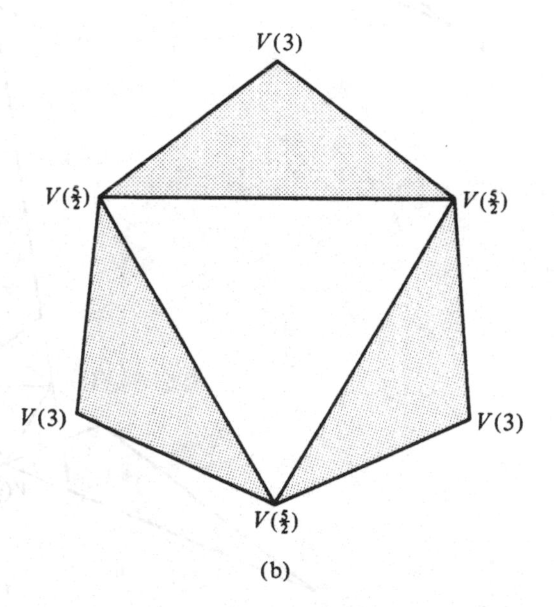

Fig. 36. Dorman Luke construction and stellation pattern for dual of **70**.

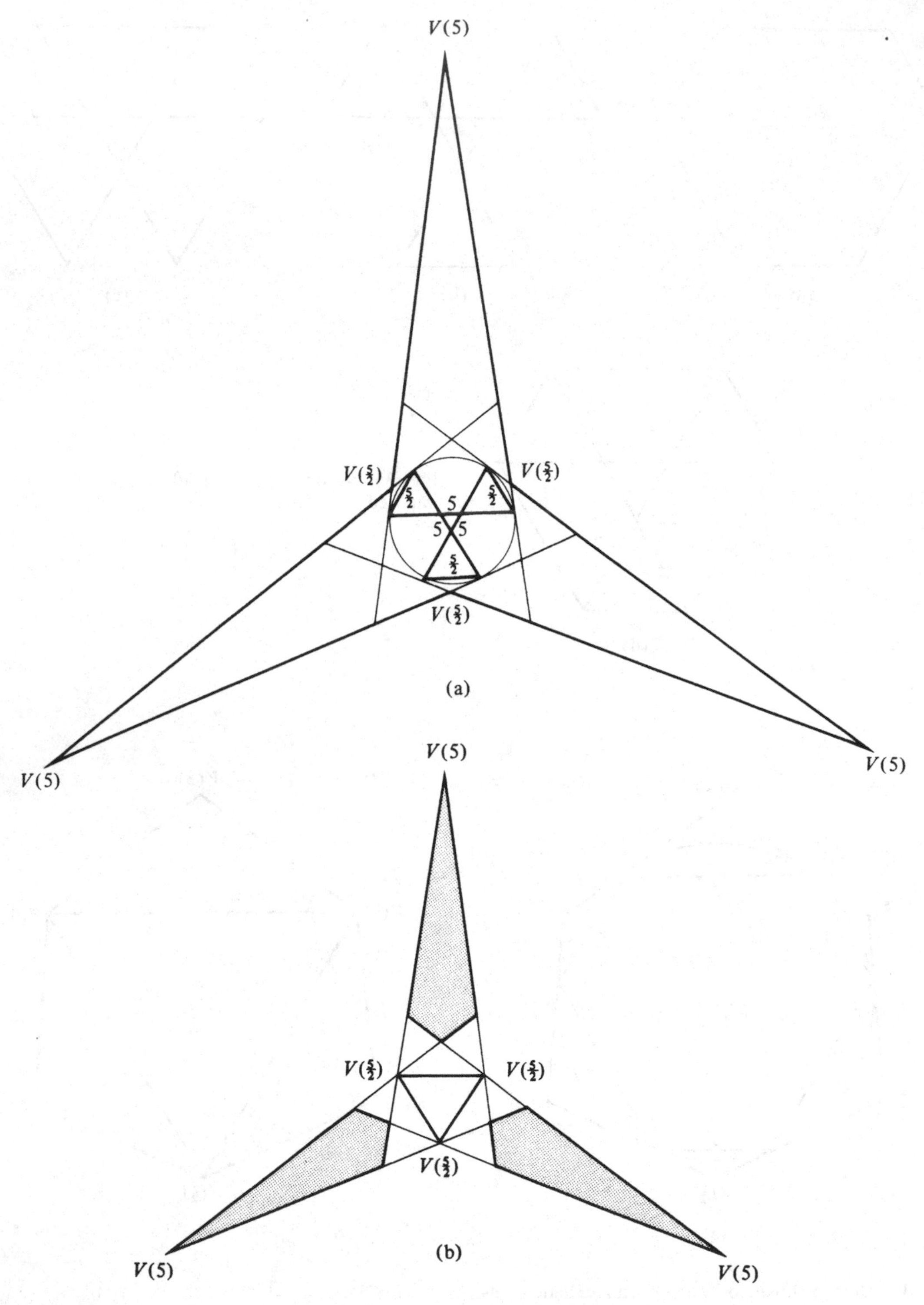

Fig. 37. Dorman Luke construction and stellation pattern for dual of **80**.

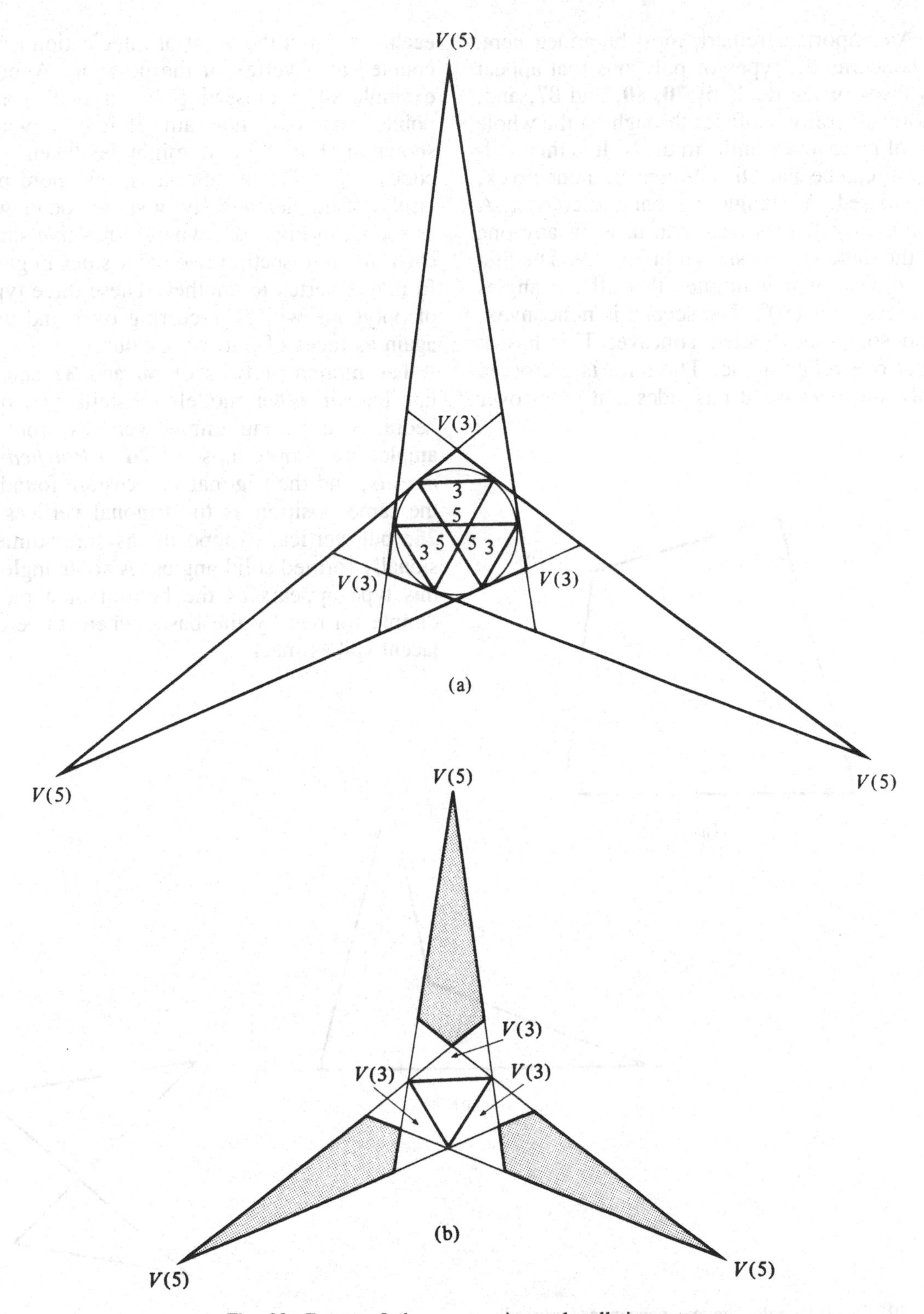

Fig. 38. Dorman Luke construction and stellation pattern for dual of **87**.

An important remark must be added here; it concerns the types of polygons that appear as faces of the duals of **70**, **80**, and **87**, and, more generally, as faces throughout the whole set of nonconvex uniform duals. It is this: Polygons can be classified as convex, nonconvex, or crossed. A triangle can only be convex. A quadrilateral, however, can take on any one of the three shapes shown in Fig. 39. The first is convex, which implies that all the angles are less than 180°. The second is nonconvex, also sometimes called concave. This has at least one reflex angle. The third is a crossed polygon, because it has sides that cross over each other, but the point of intersection is not counted as a vertex of the polygon. Another example of a crossed polygon is the five-pointed star or pentagram. If it is drawn as shown in Fig. 40a, it might be taken as a concave polygon of ten sides; but more precisely, or mathematically, it should be viewed as shown in Fig. 40b, where it has five sides, each side intersecting two other sides in going from one vertex to another. These three types of polygons will be recurring over and over again as faces of nonconvex duals.

The hidden vertices of **80** and **87** can be laid bare in other models of stellated icosahedra. The pentagrammic vertices, for example, are simply those of **26** in *Polyhedron models*, and the trigonal vertices are found in the same position as the trigonal vertices of **26**, but vertically opposite as three-dimensionally formed solid angles. A solid angle of this type appears as the bottom of a pit or dimple formed by the bases where three adjacent spikes meet.

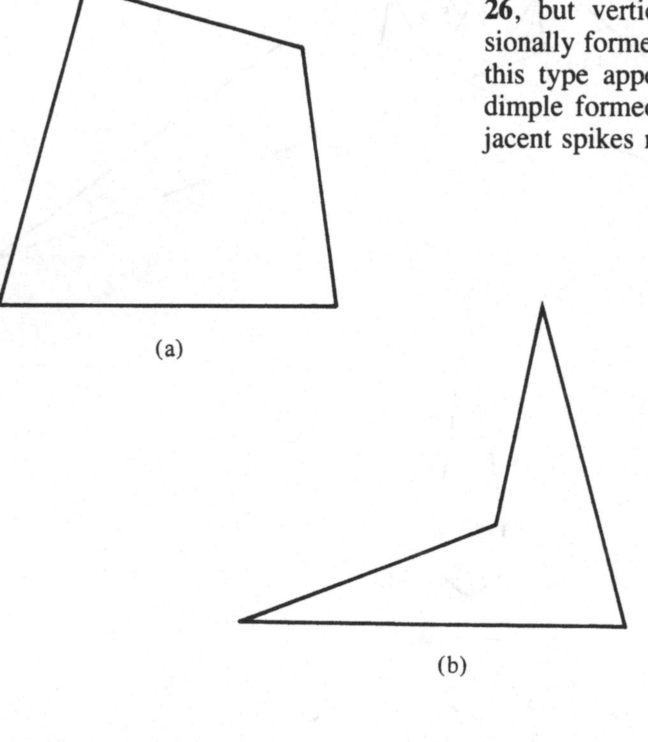

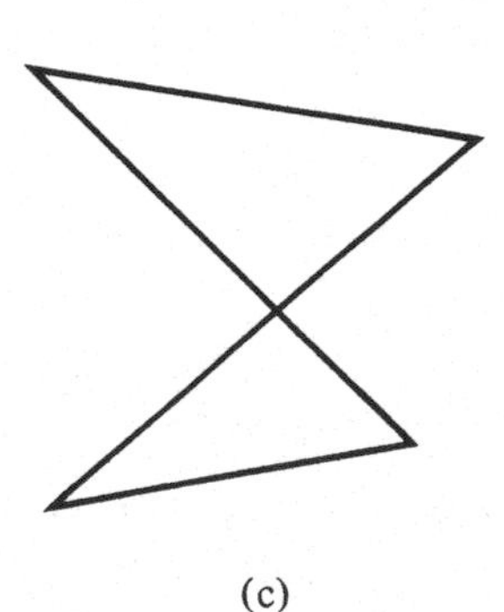

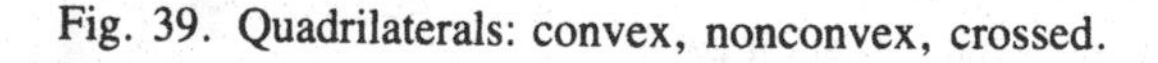

Fig. 39. Quadrilaterals: convex, nonconvex, crossed.

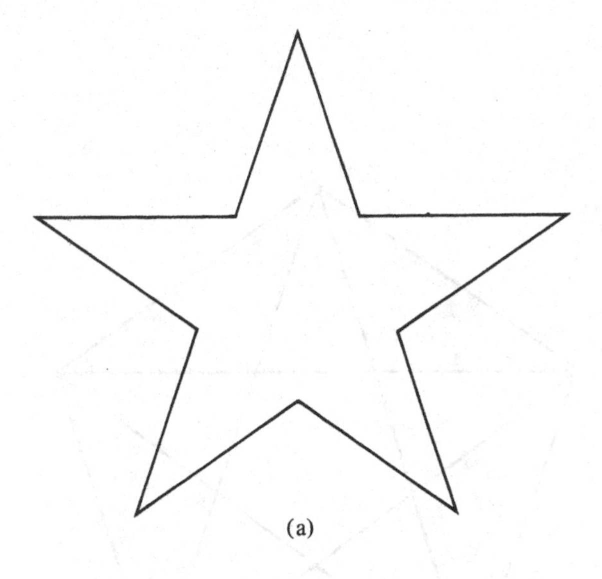

(a)

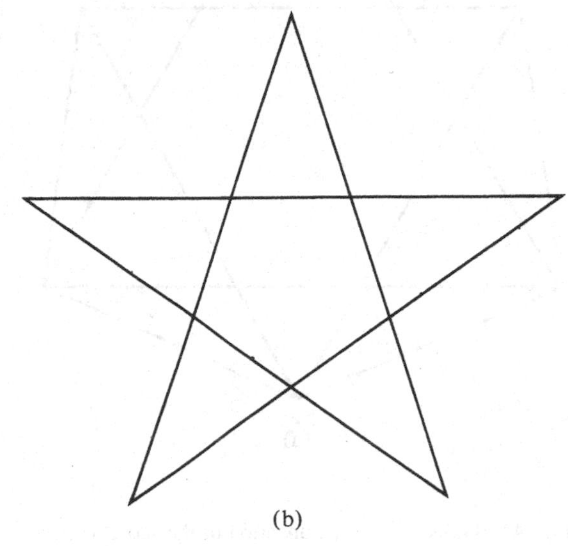

(b)

Fig. 40. Pentagram or five-pointed star.

The pentagrammic vertices of **26** can be made most visibly explicit in the special model shown in Photo 30. This is one of those inviting side paths along the way. This model is actually a compound of a regular dodecahedron interpenetrating the first stellation of the icosahedron. It can be derived as a stellated form of a truncated icosahedron, but the truncation is not the one that yields the Archimedean form. It may be called a golden-section truncation of the icosahedron. Figure 41 shows the two faces involved, with shaded portions showing the exterior parts. Chilton says this form is one of the principal shadows of the four-dimensional polytope (5, 3, 5/2). It makes an attractive model. Further stellations of this will yield other compounds of the dodecahedron interpenetrating stellated forms of the icosahedron. For example, Photo 31 shows a

Photo 30. Compound of dodeca and icosa stellations.

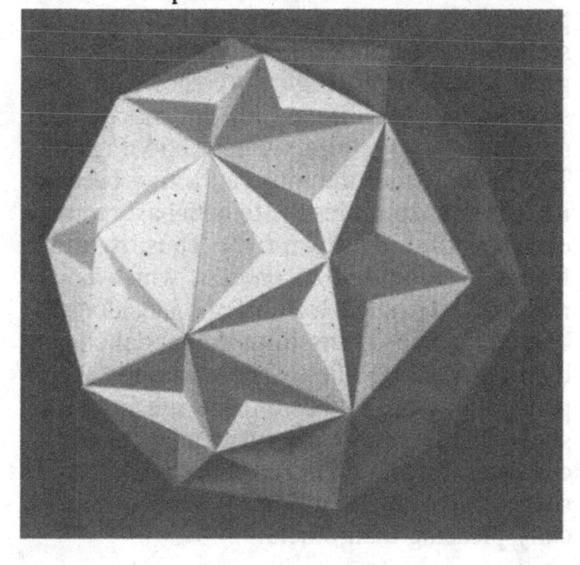

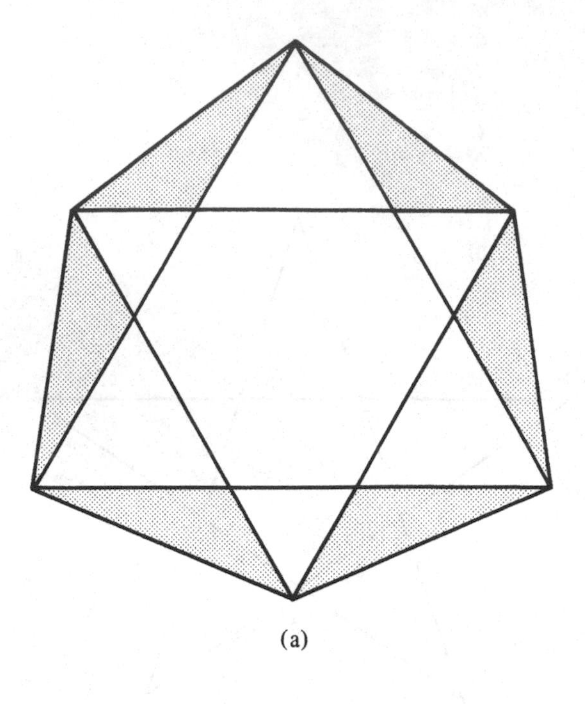

(a)

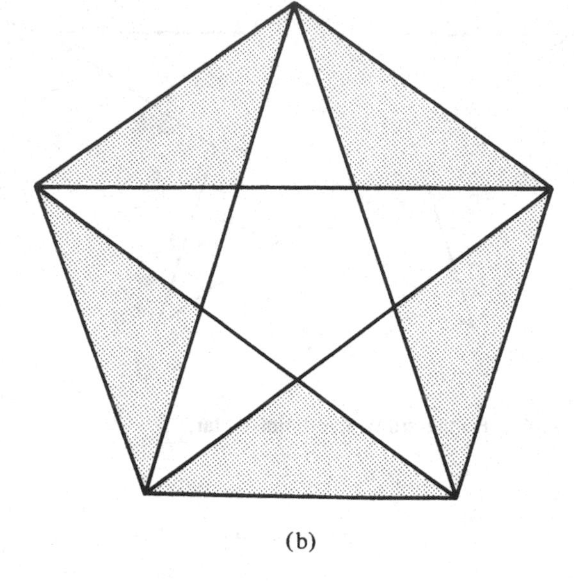

(b)

Fig. 41. Golden-section truncation of the icosahedron.

particularly attractive compound of the great stellated dodecahedron interpenetrating the great icosahedron. Figure 42 shows in dark shading the parts needed for this model. The stippled shading relates Fig. 41 to Fig. 42. This model is all the more attractive because it gives equal prominence to both basic forms, each beautiful in its own right. Thus, it differs from **61** in *Polyhedron models*, which is derived as a stellated form of the icosidodecahedron. It also differs from the model shown by Cundy and Rollett. These various interpenetrations of two basic forms are instances showing how variations of Archimedean forms can often lead to different but closely related shapes, some quite obviously more aesthetically pleasing than others.

Photo 31. Compound of great stellated dodeca and great icosa.

Fig. 42. Extended pattern for golden-section truncation of the icosahedron: (a) the pentagon plane; (b) the hexagon plane.

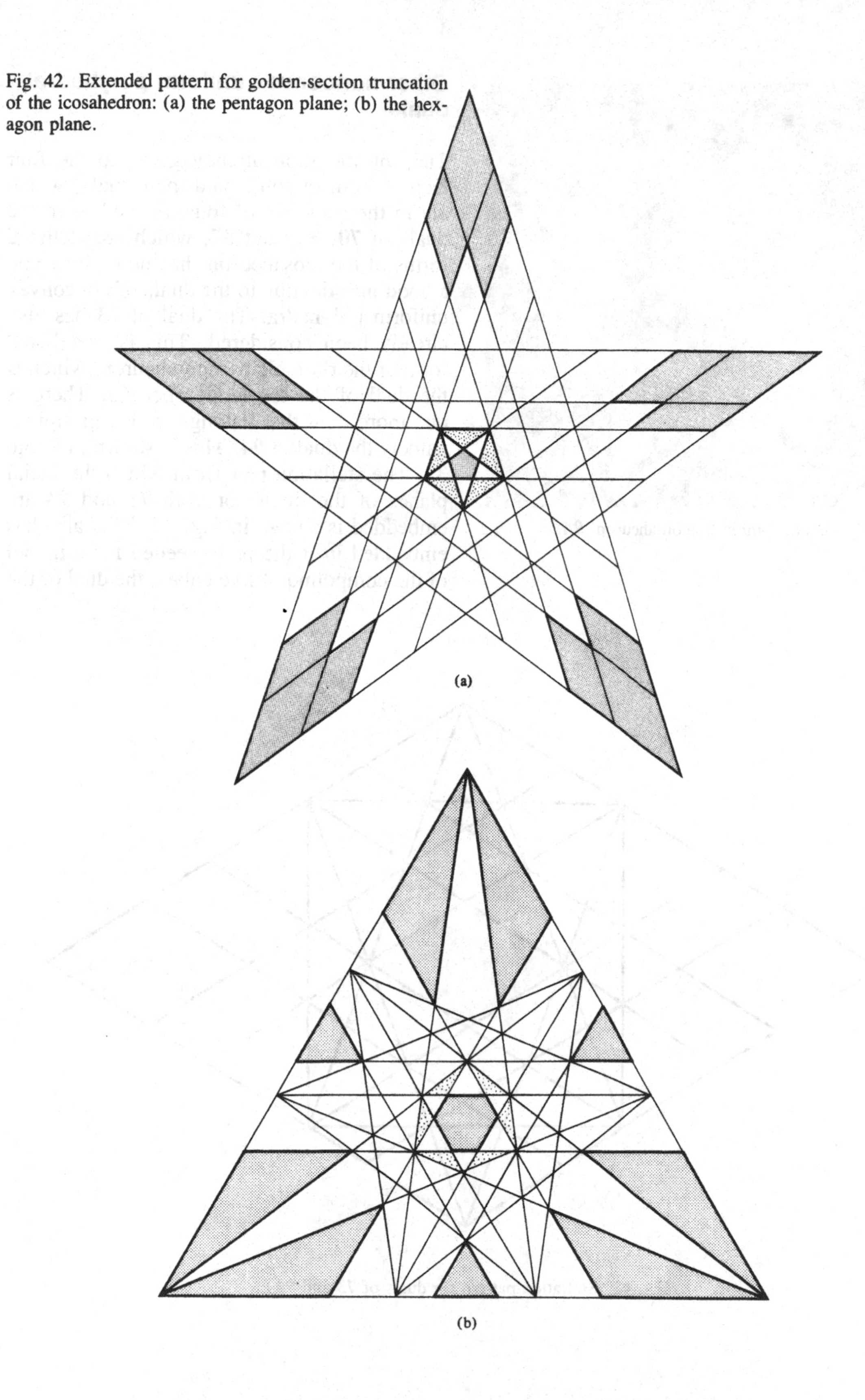

Photo 32. Great rhombic triacontahedron (**94**).

Other nonconvex uniform polyhedral duals

The consideration already given to the four Kepler–Poinsot solids and their duals, which are in the same set of four, as well as to the duals of **70**, **80**, and **87**, which are stellated forms of the icosahedron, has now given you a good introduction to the duals of nonconvex uniform polyhedra. The dual of **73** has also already been considered. This is a stellated form of the rhombic triacontahedron, which is the dual of the icosidodecahedron. There is one more dual that belongs to this grouping, namely the dual of **94**. This is shown in Photo 32. The stellation pattern in which the facial planes of the duals for both **73** and **94** are embedded is shown in Fig. 43. This also has embedded in it the parts needed for a model of the compound of five cubes, the dual of the

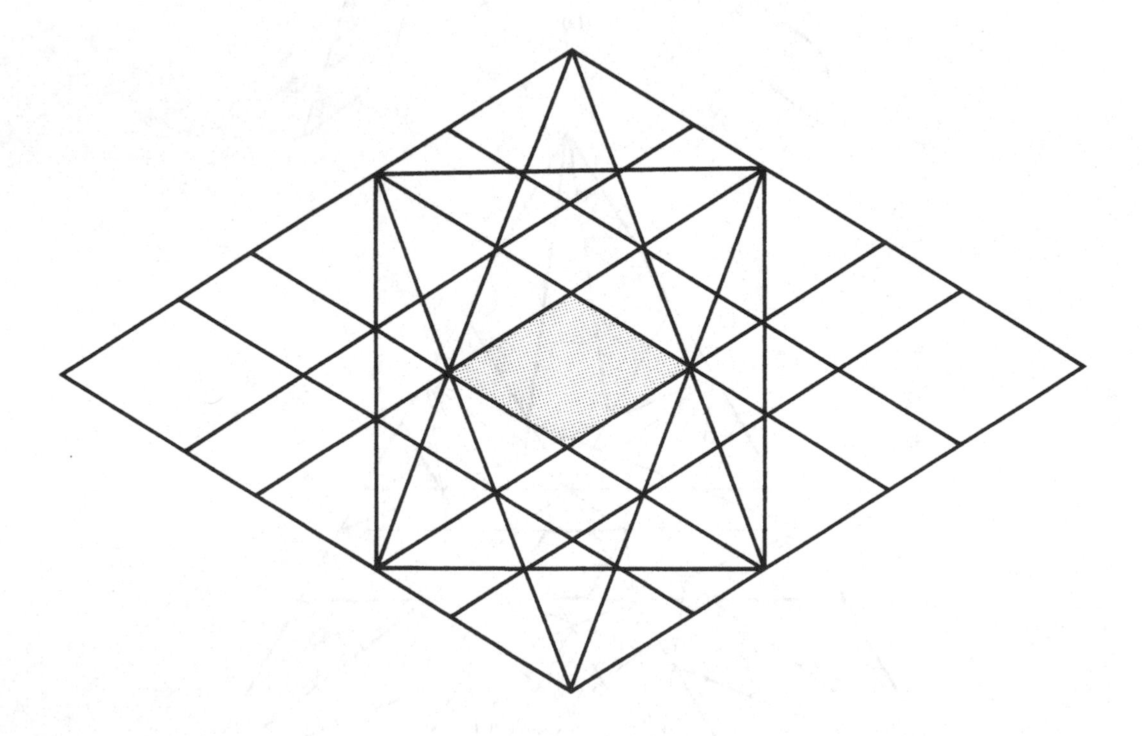

Fig. 43. Stellation pattern for duals of **73** and **94**.

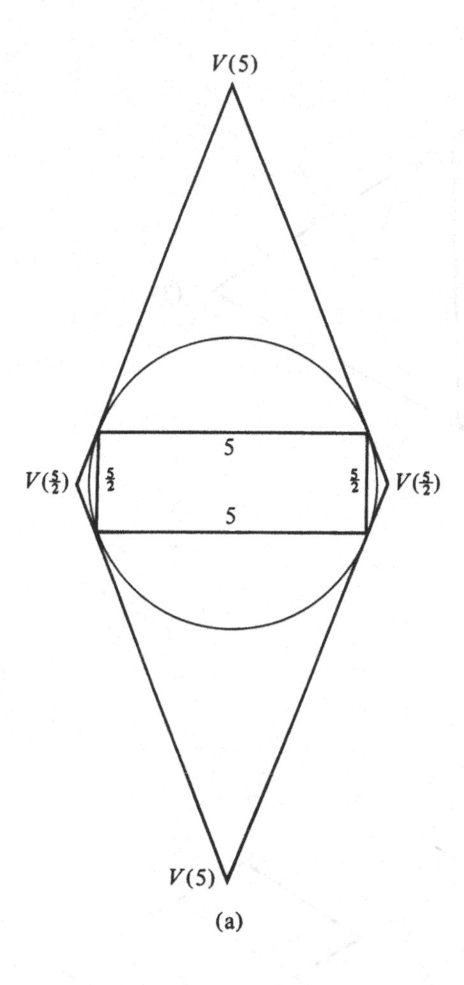

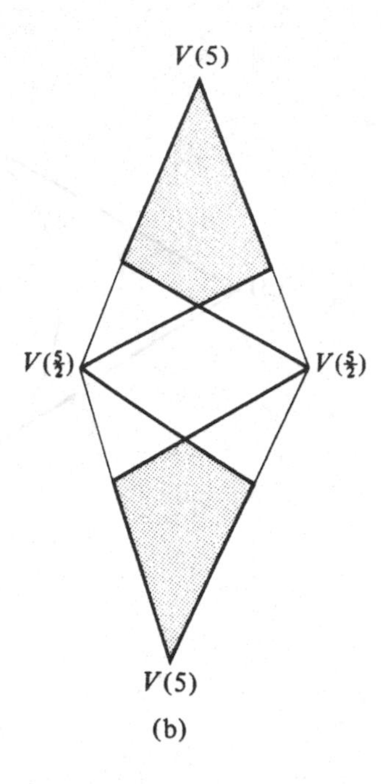

Fig. 44. Dorman Luke construction and stellation pattern for dual of **73**.

compound of five octahedra. This latter compound is a stellated form of the icosahedron. Notice the richness of interrelationships here. Figures 44 and 45 show the Dorman Luke construction derived from respective vertex figures from which these facial planes are also derived. These duals are well known to anyone acquainted with the work of Cundy and Rollett (*Mathematical models*) and Coxeter (*Regular polytopes*).

Duals derived from other Archimedean forms

A careful examination of other nonconvex uniform polyhedra will reveal that some of them are obviously related to convex Archimedean forms. It was this fact that led Badoureau, in the late nineteenth century, to discover as many as thirty-seven nonconvex uniform polyhedra, many of them unknown until his day. So, in working out the duals of these, all that is needed is some acquaintance with the duals of the convex Archimedean forms and the stellation process. Furthermore, the fact that the Dorman Luke construction still holds for nonconvex forms will make it possible to examine vertex figures in each of these instances and from them to derive the shapes of facial planes for these duals. It is then usually not at all difficult to see where these facial planes are embedded in the stellation pattern. So, without further comment, you are invited to examine Photos 33 through 42 and the accompanying drawings shown in Figs. 46 through 61 in connection with the duals listed in Table 2. In this table, the grouping in sets of three is derived from

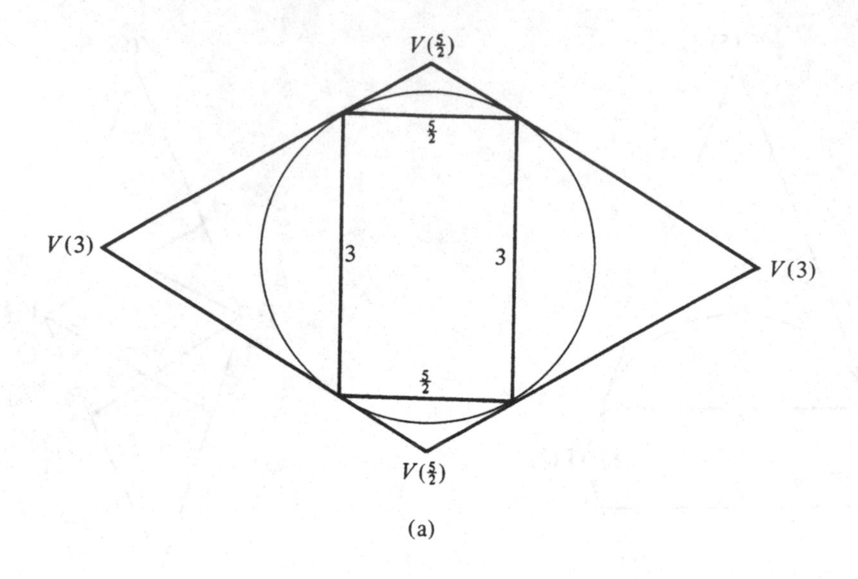

(a)

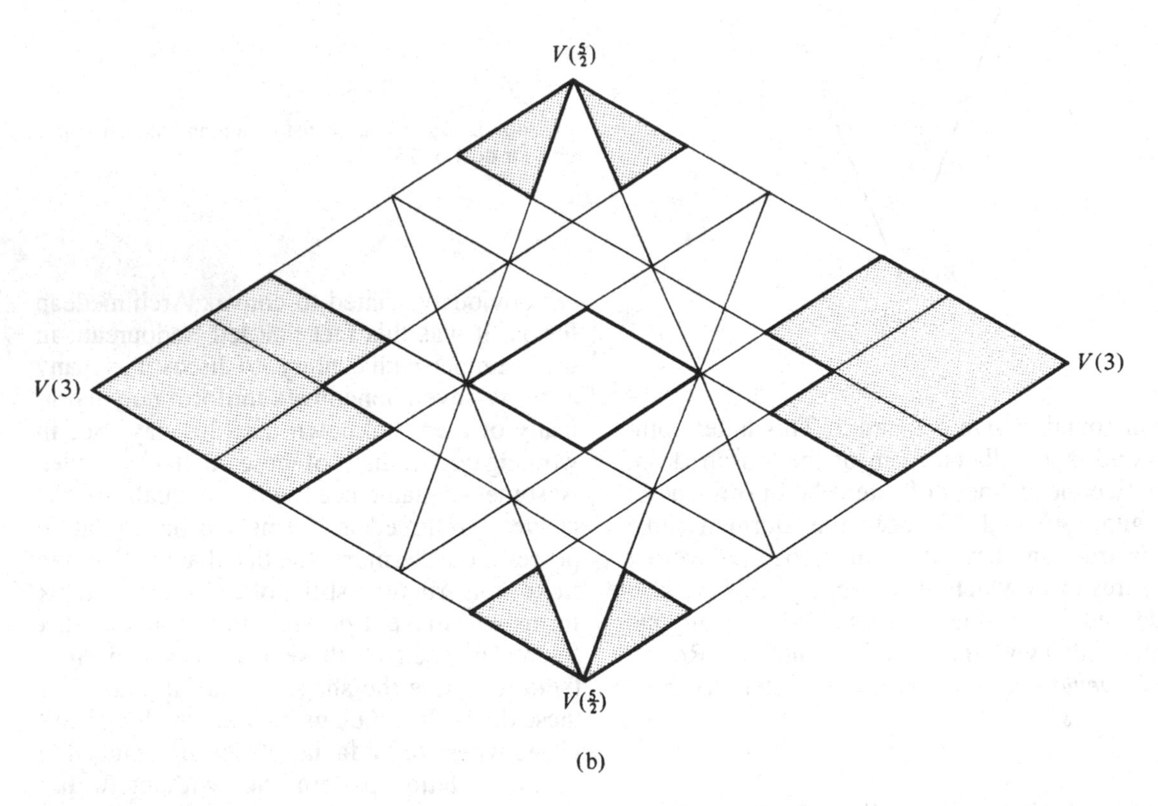

(b)

Fig. 45. Dorman Luke construction and stellation pattern for dual of **94**.

the fact that each set of three has the same convex hull, and hence the vertex figures are closely related. The shapes of faces for these duals also portray the various types of polygons mentioned previously: convex, nonconvex, and crossed. In fact, each set of three shows one instance of each. When the same photo is named as the dual of two different original polyhedra, you may remember that this is due to the placement of hidden vertices.

Table 2. *Duals derived from Archimedean forms*

Polyhedron	Stellation pattern	Convex hull	Dual of hull	Dual
69	Fig. 46	Rhombicubocta	Deltoidal	Photo 33
86			icositetra	Photo 33
92				Photo 34
77	Fig. 50	Truncated hexa	Triakisocta	Photo 35
85				Photo 36
103				Photo 37
81	Fig. 54	Truncated dodeca	Triakisicosa	Photo 38
88				Photo 39
101				Photo 40
72	Fig. 58	Rhombicosidodeca	Deltoidal	Photo 41
74			hexeconta	Photo 41
97				Photo 42

Photo 33. Small hexacronic icositetrahedron (**69**) and small rhombihexacron (**86**).

Photo 34. Great triakisoctahedron (**92**).

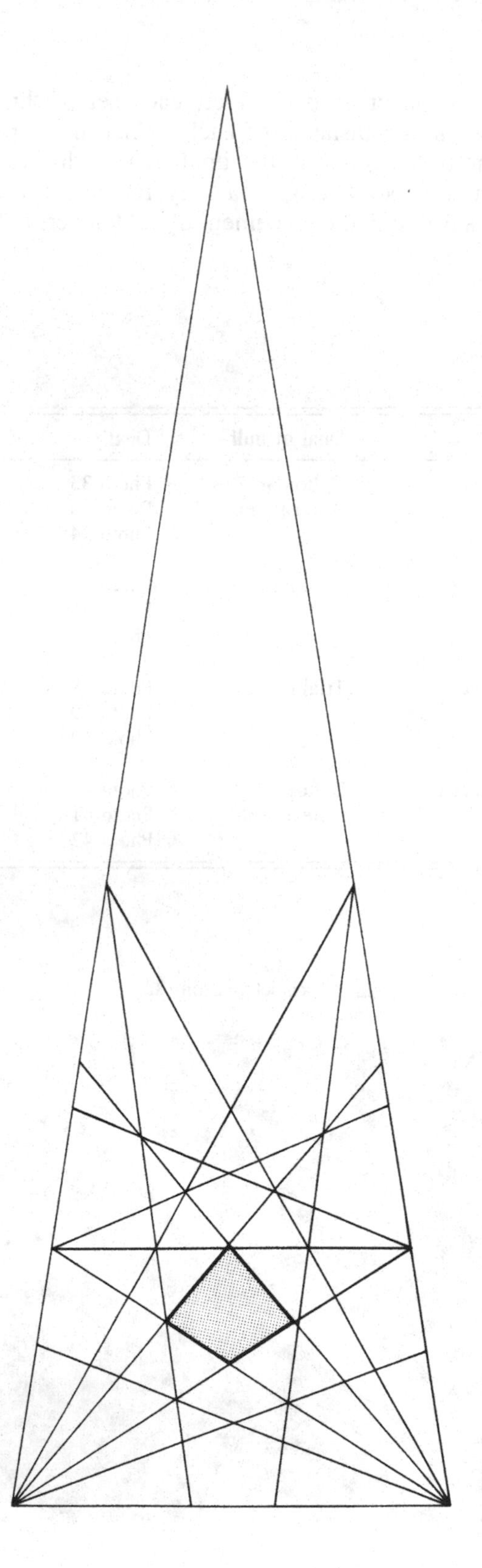

Photo 35. Great hexacronic icositetrahedron (**77**).

Fig. 46. Partial stellation pattern for duals of **69**, **86**, and **92**.

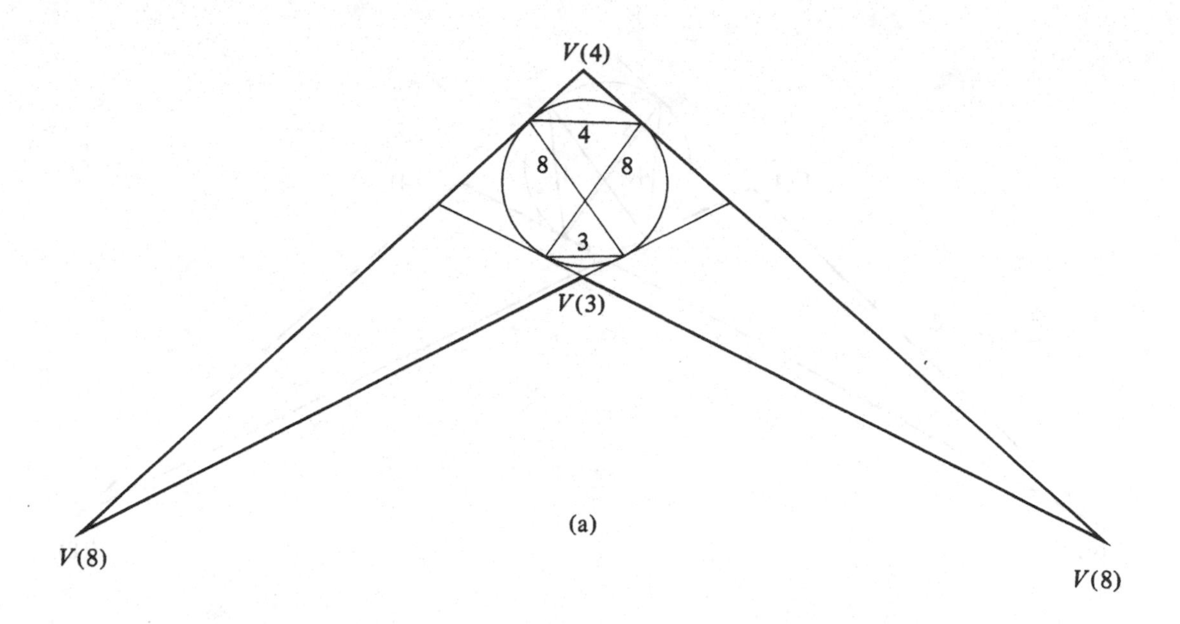

Fig. 47. Dorman Luke construction and stellation pattern for dual of **69**.

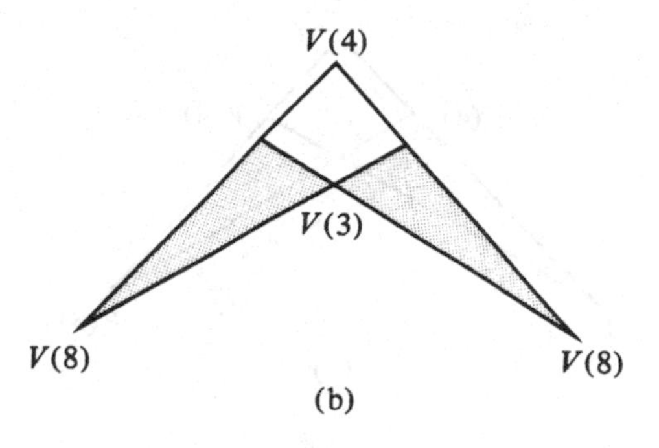

Photo 36. Great deltoidal icositetrahedron (**85**).

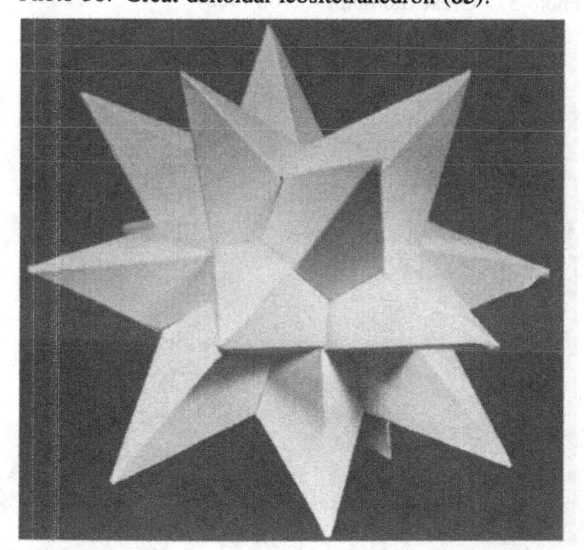

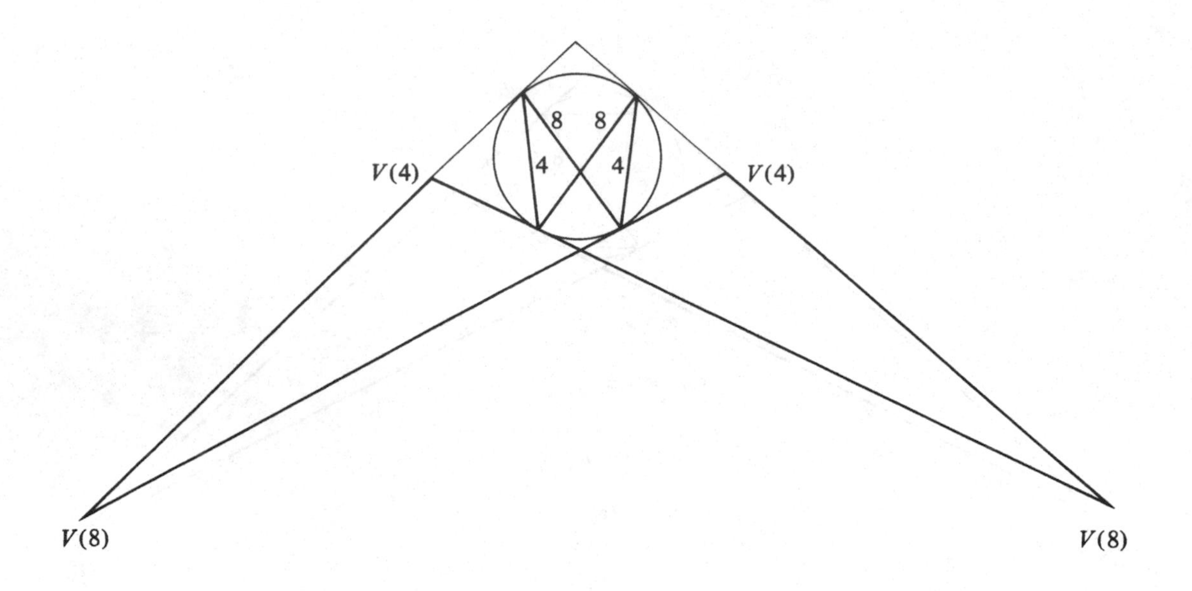

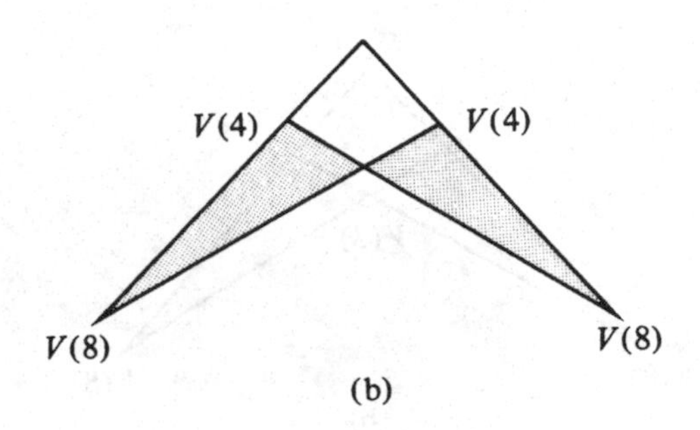

Fig. 48. Dorman Luke construction and stellation pattern for dual of **86**.

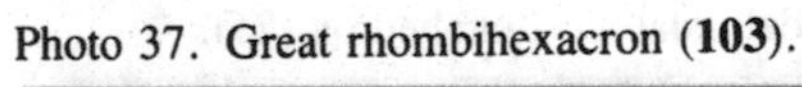

Photo 37. Great rhombihexacron (**103**).

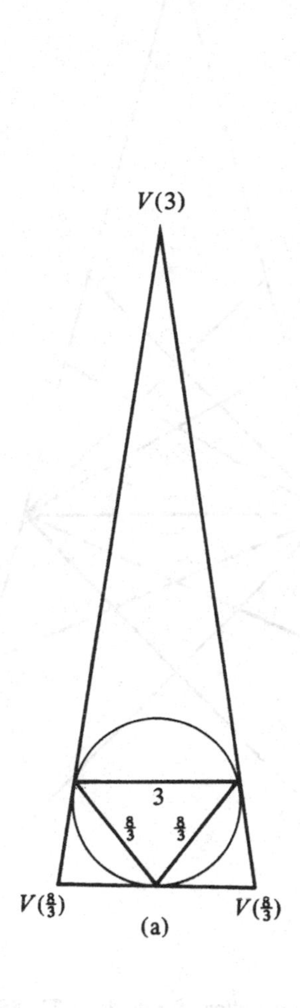

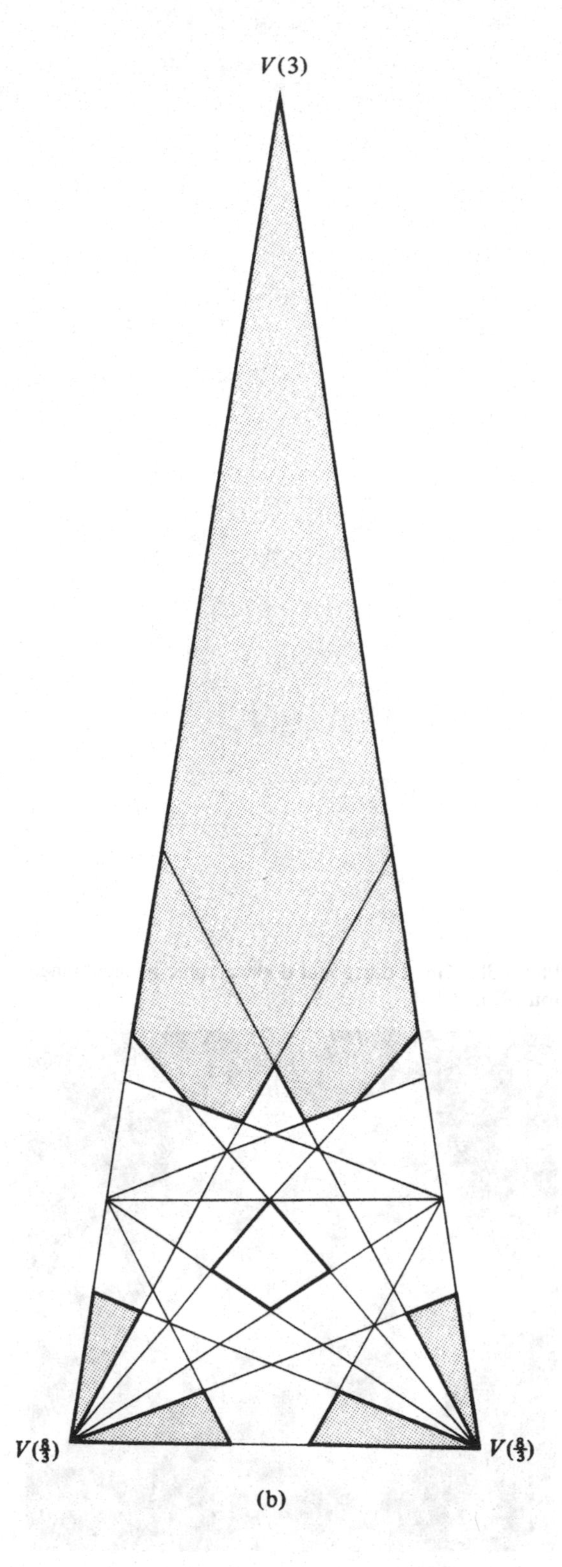

Fig. 49. Dorman Luke construction and stellation pattern for dual of **92**.

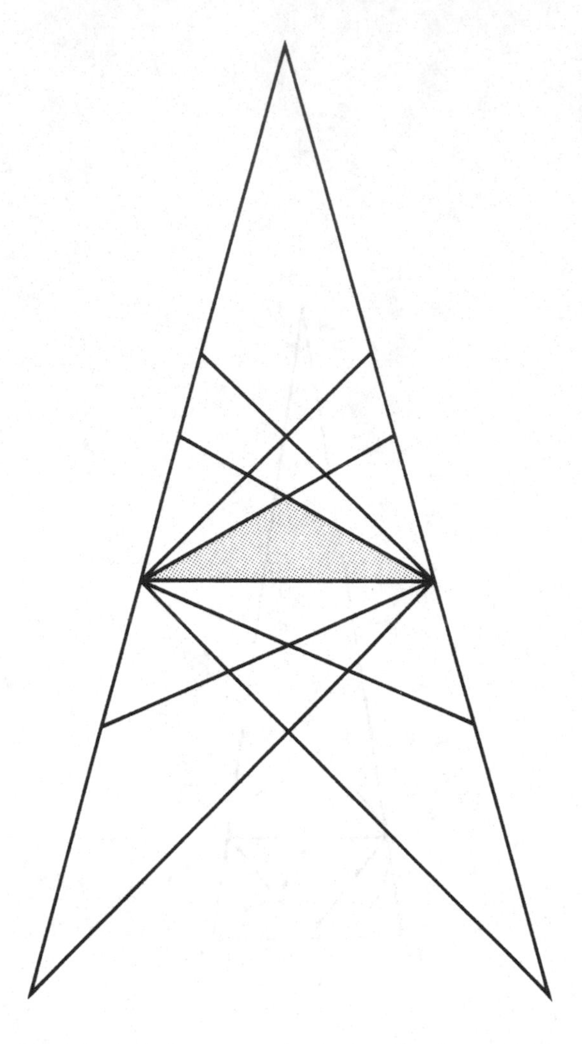

Fig. 50. Stellation pattern for duals of **77**, **85**, and **103**.

Photo 38. Great ditrigonal dodecacronic hexecontahedron (**81**).

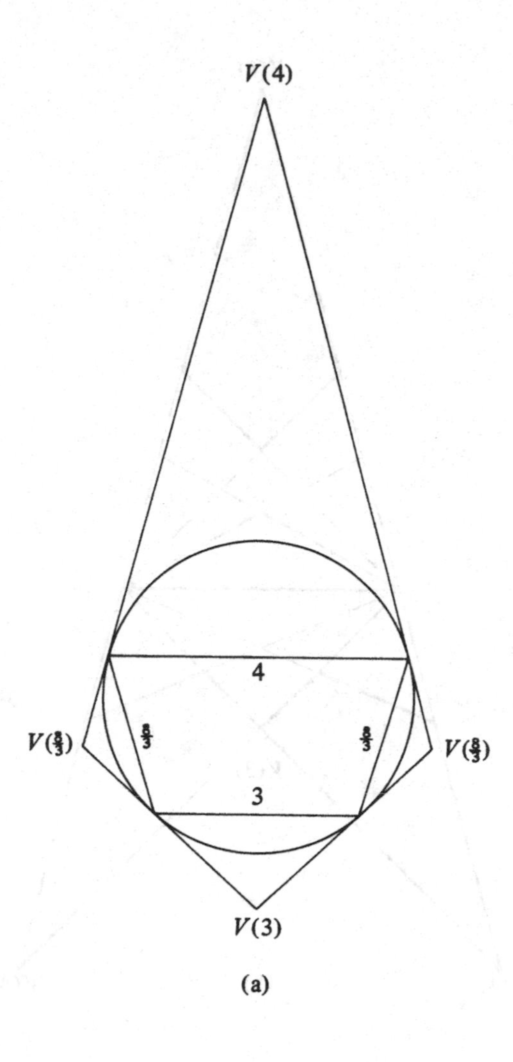

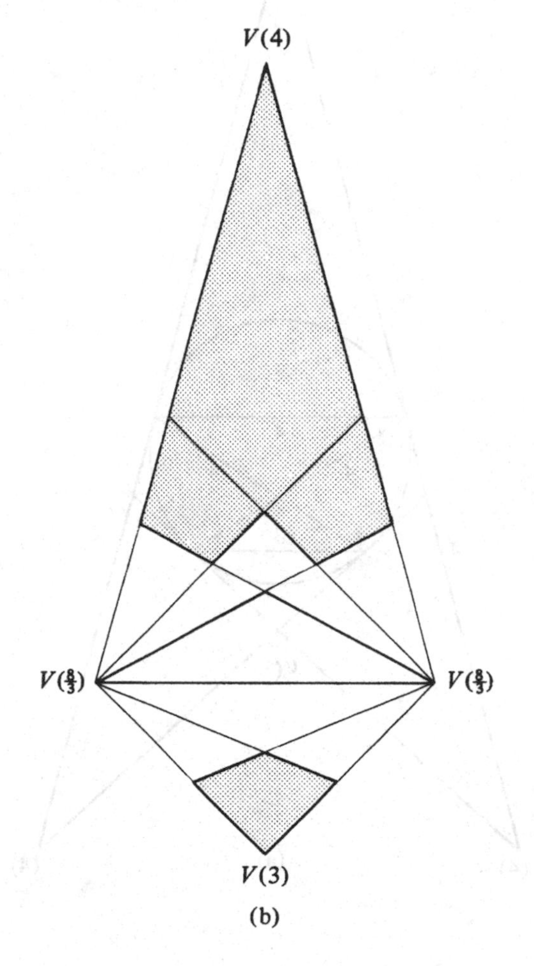

Fig. 51. Dorman Luke construction and stellation pattern for dual of **77**.

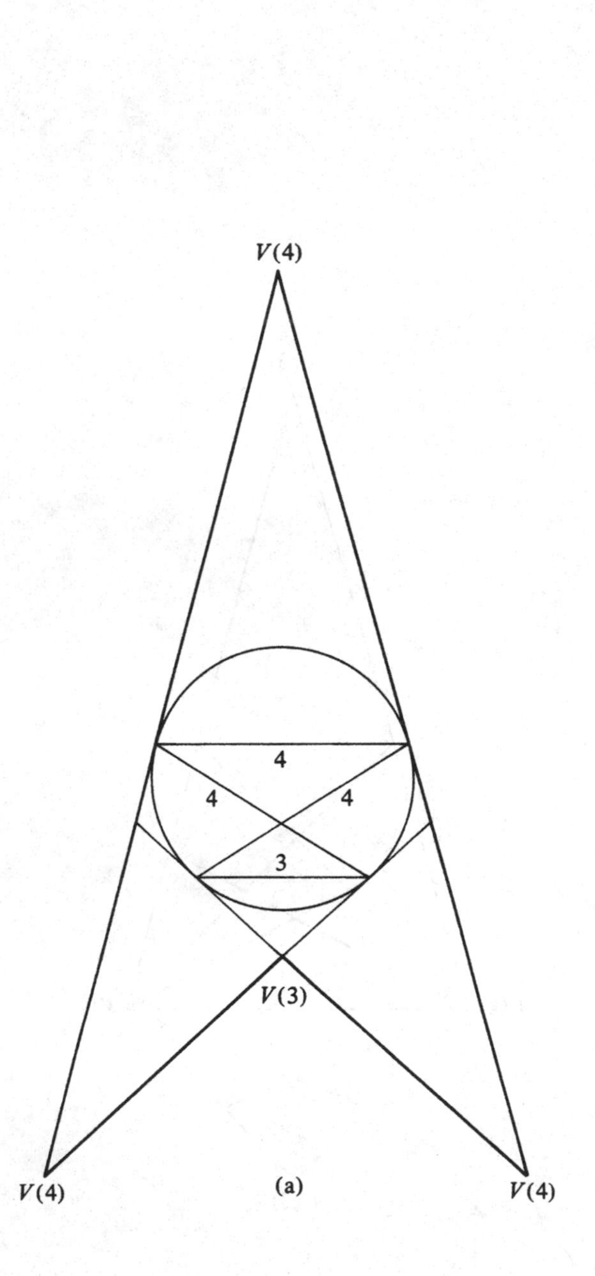

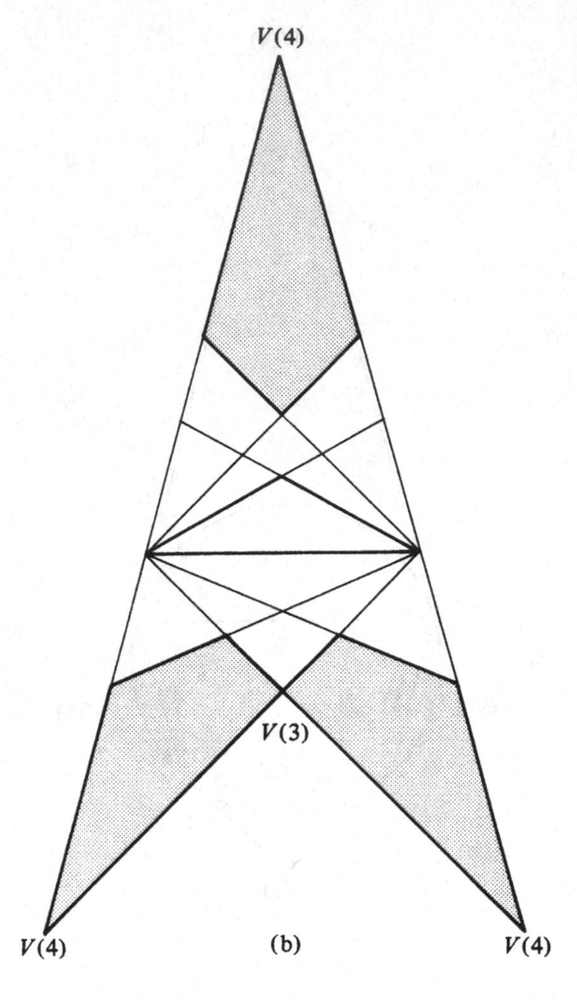

Fig. 52. Dorman Luke construction and stellation pattern for dual of **85**.

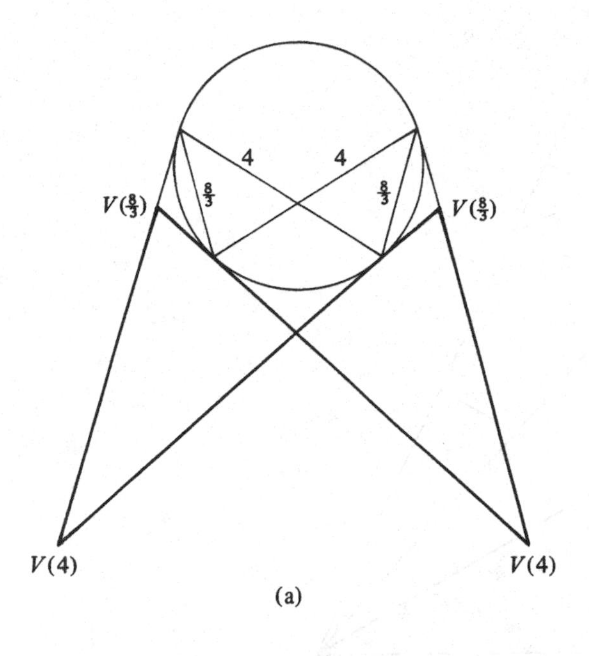

Photo 39. Great icosacronic hexecontahedron (**88**).

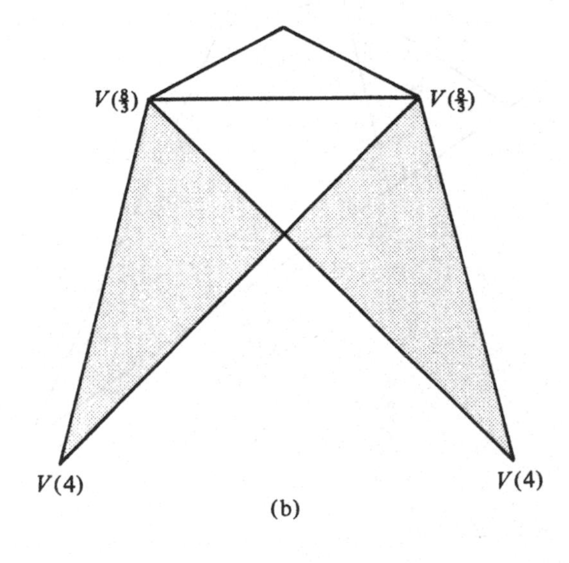

Fig. 53. Dorman Luke construction and stellation pattern for dual of **103**.

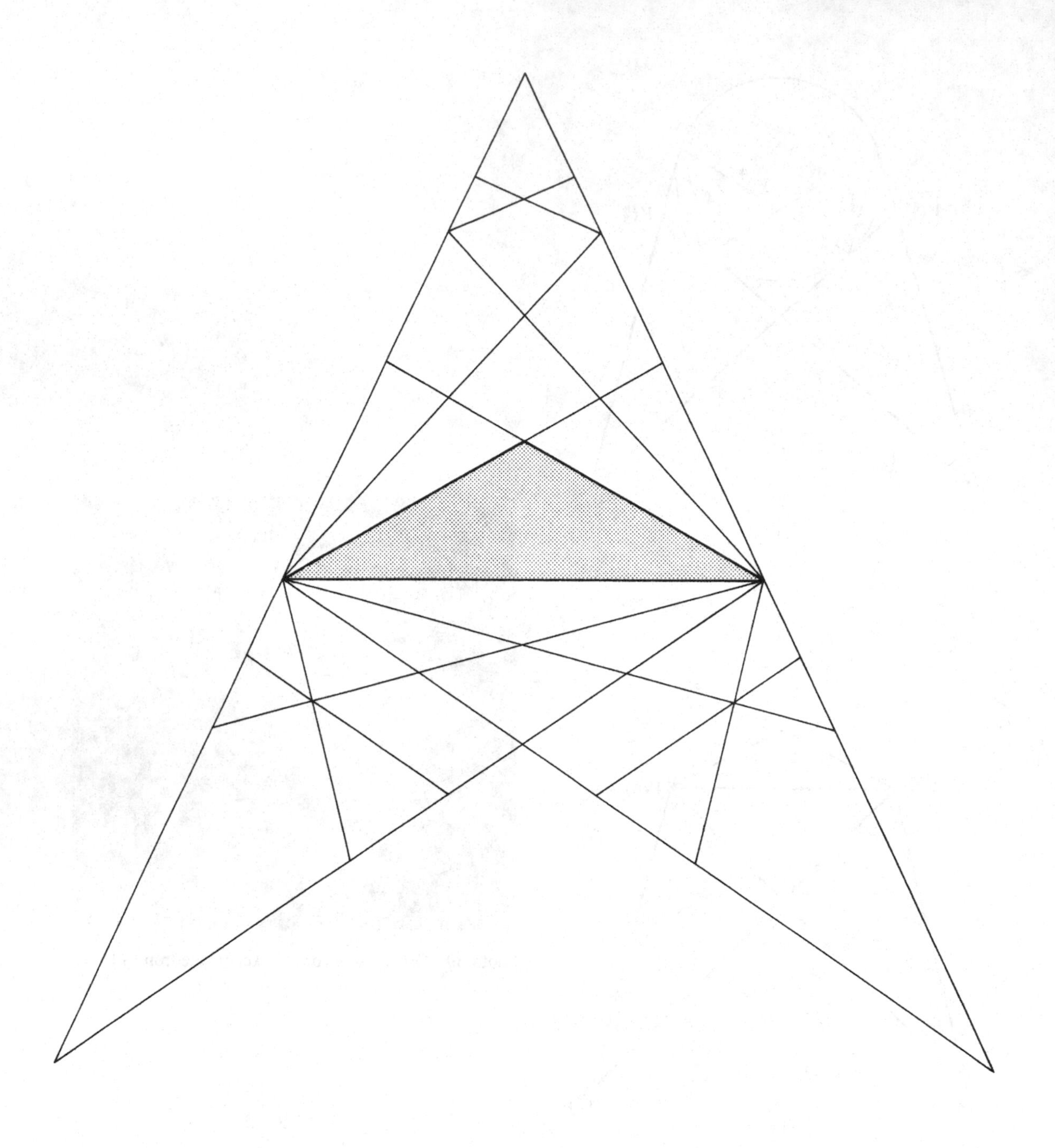

Fig. 54. Stellation pattern for duals of **81**, **88**, and **101**.

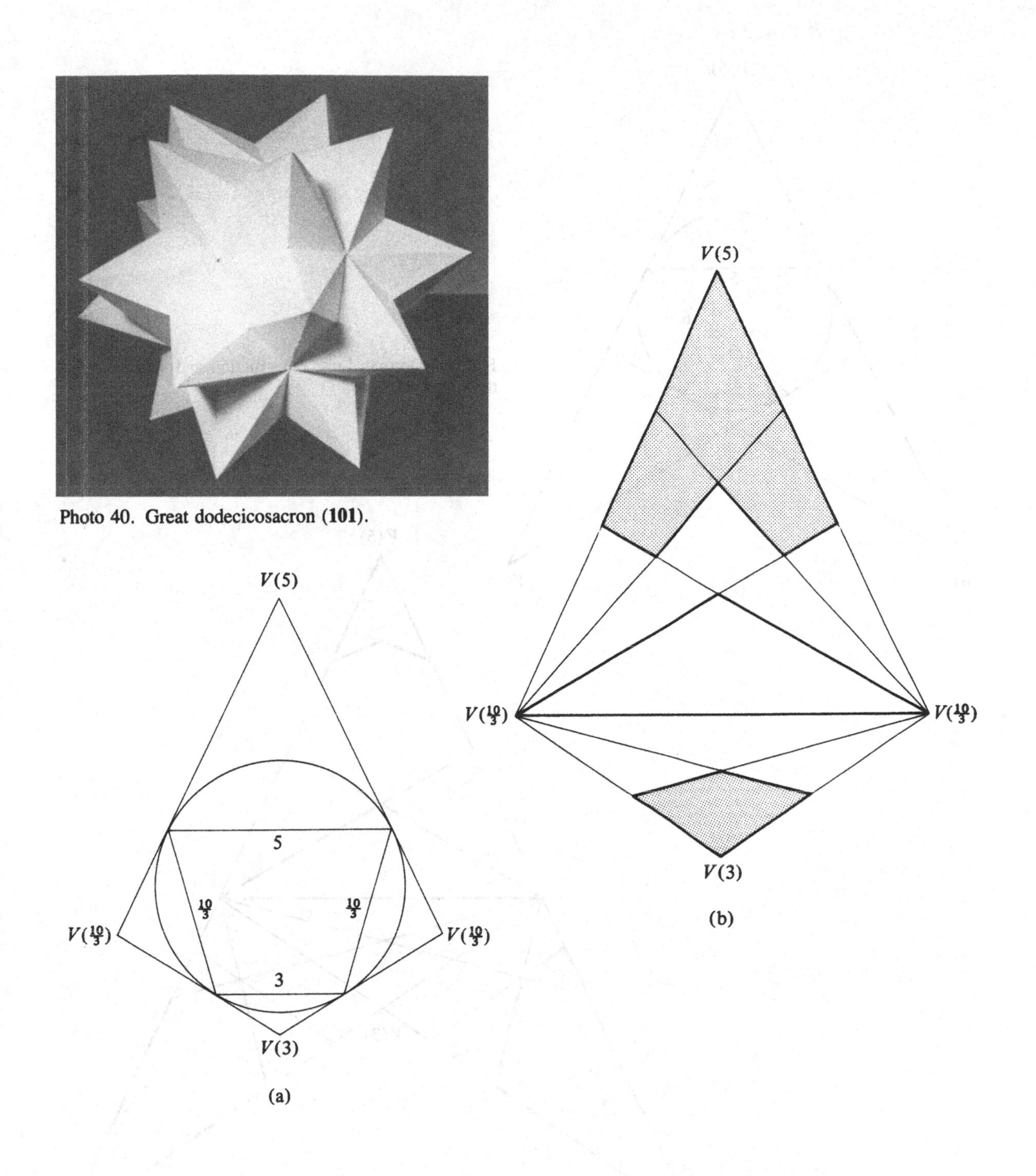

Photo 40. Great dodecicosacron (**101**).

Fig. 55. Dorman Luke construction and stellation pattern for dual of **81**.

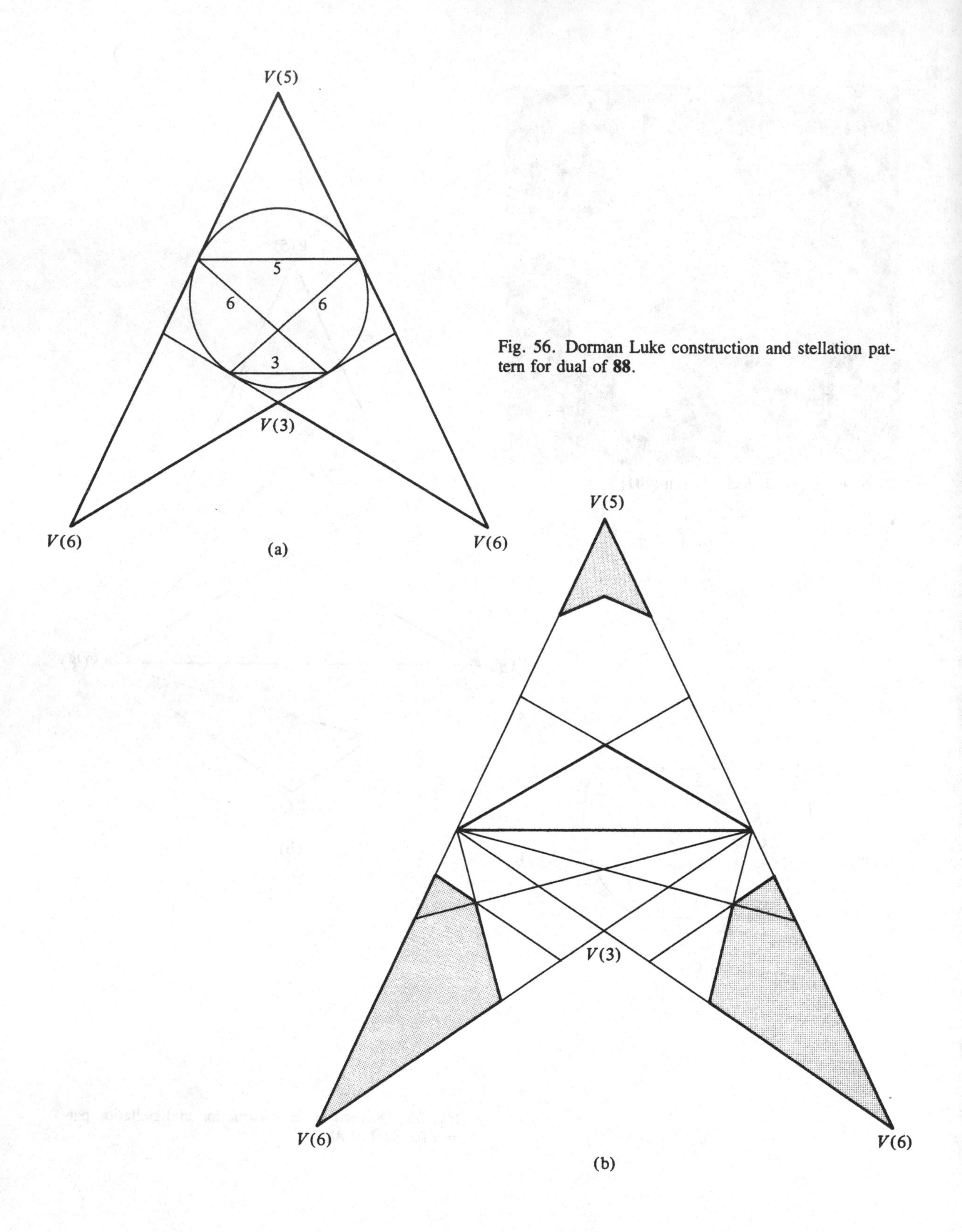

Fig. 56. Dorman Luke construction and stellation pattern for dual of **88**.

Fig. 57. Dorman Luke construction and stellation pattern for dual of **101**.

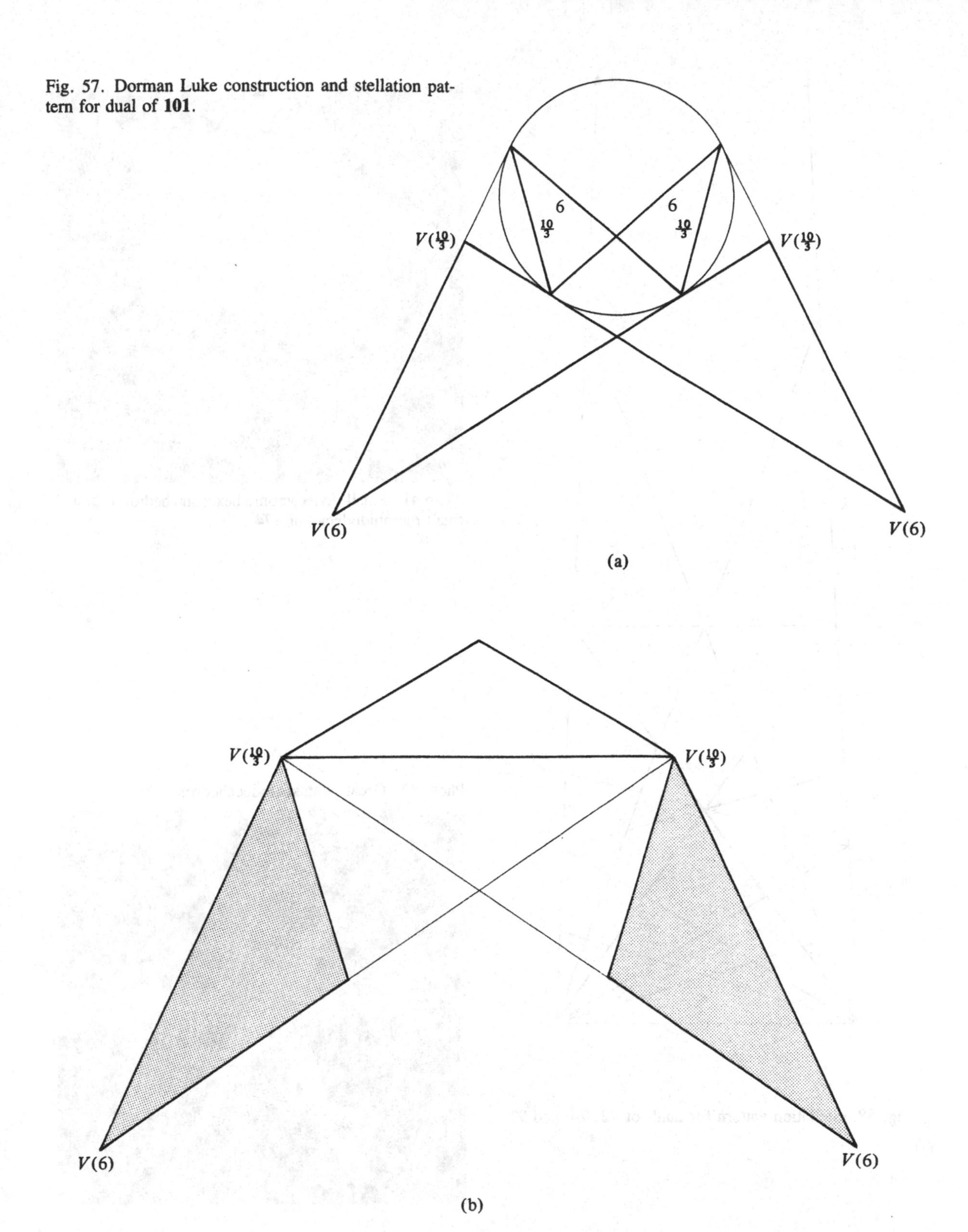

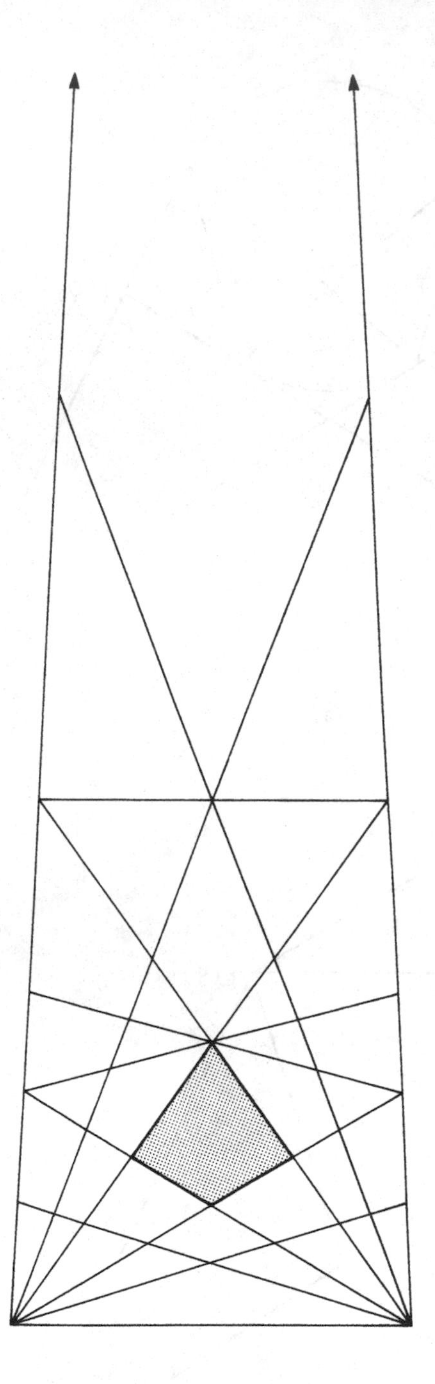

Fig. 58. Stellation pattern for duals of **72**, **74**, and **97**.

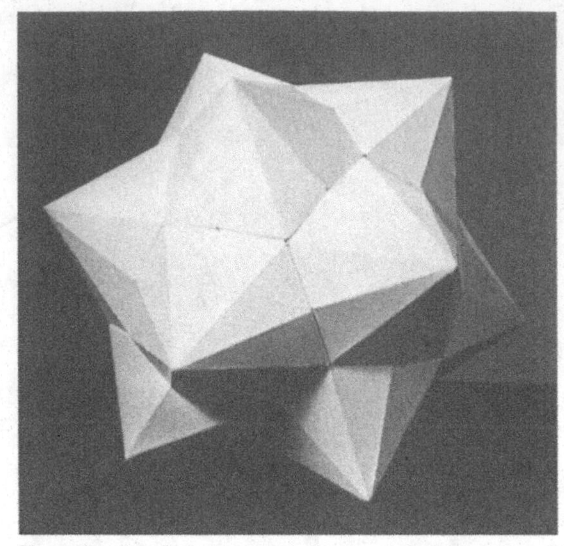

Photo 41. Small dodecacronic hexecontahedron (**72**) and small rhombidodecacron (**74**).

Photo 42. Great pentakisdodecahedron (**97**).

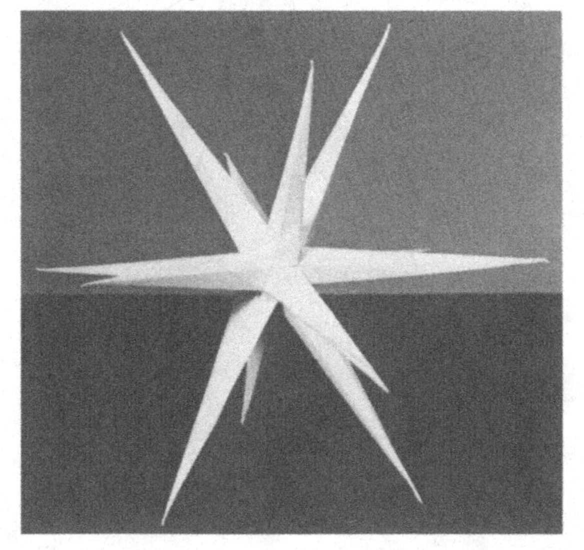

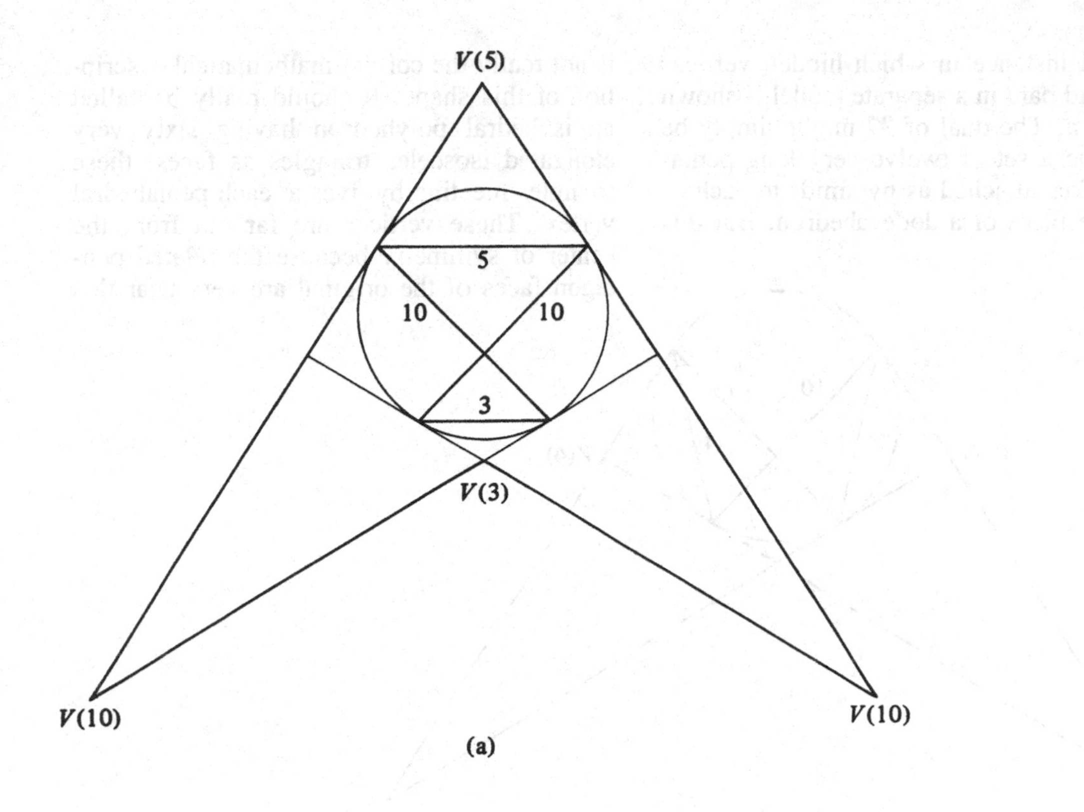

Fig. 59. Dorman Luke construction and stellation pattern for dual of **72**.

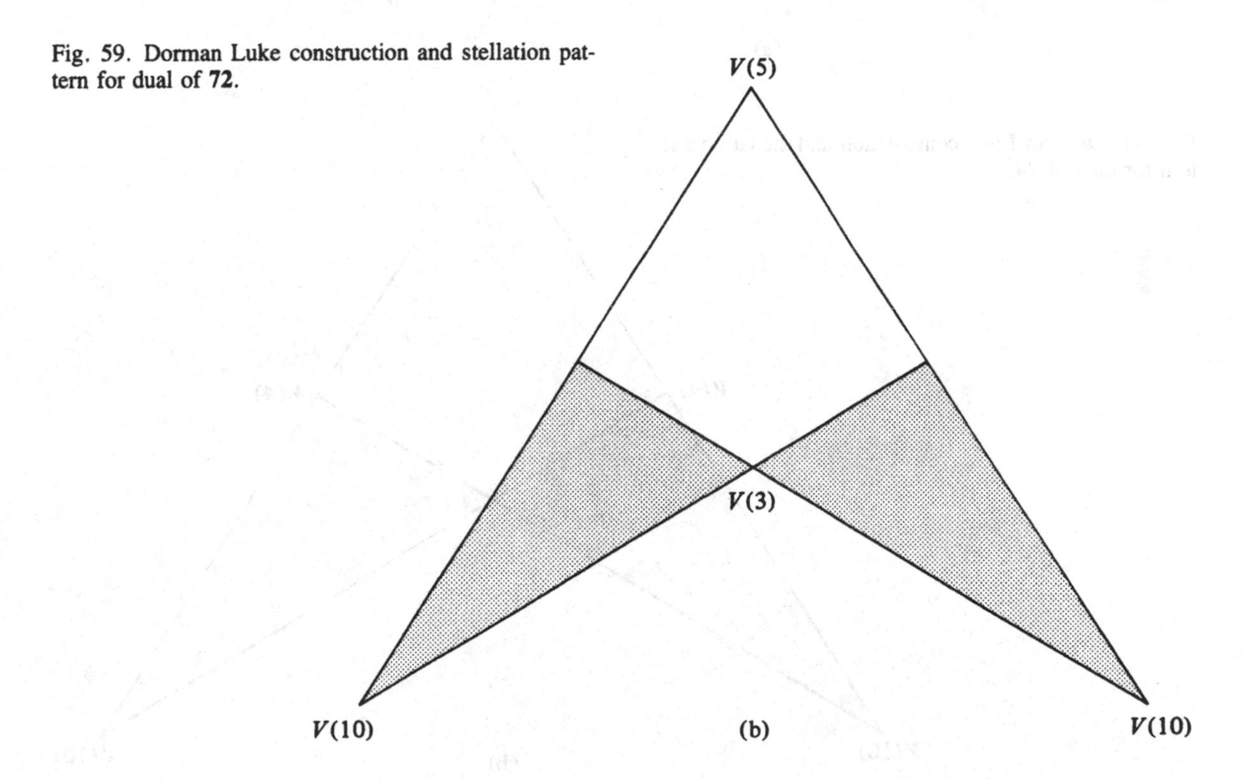

A good instance in which hidden vertices may be laid bare in a separate model is shown in Fig. 61b. The dual of **97** might simply be taken to be a set of twelve very long pentahedral spikes attached as pyramids to each of the twelve faces of a dodecahedron. But this is not really the correct mathematical description of this shape. It should really be called an isohedral polyhedron having sixty very elongated isosceles triangles as faces, these triangles meeting by fives at each pentahedral vertex. These vertices are far out from the center of symmetry because the related pentagon faces of the original are very near this

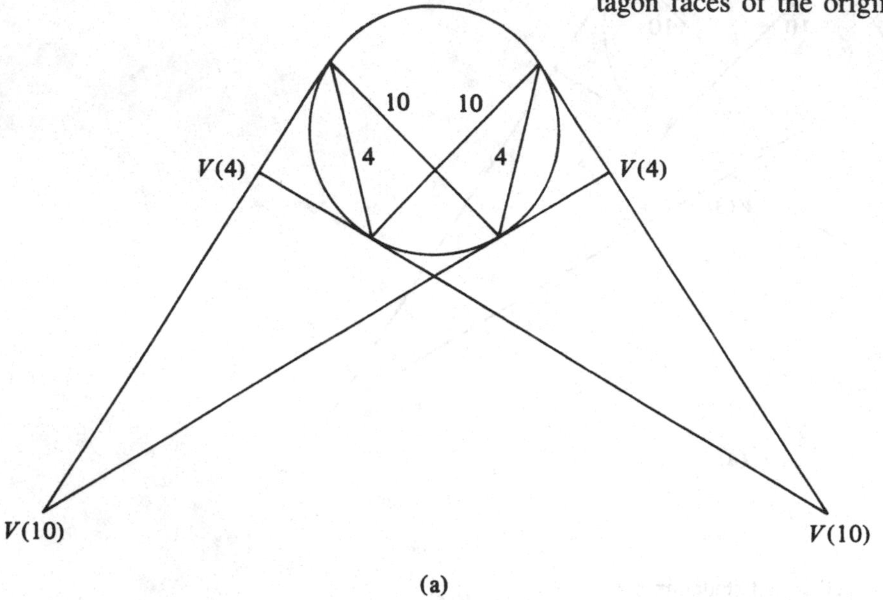

(a)

Fig. 60. Dorman Luke construction and stellation pattern for dual of **74**.

V(4) V(4)

V(10) (b) V(10)

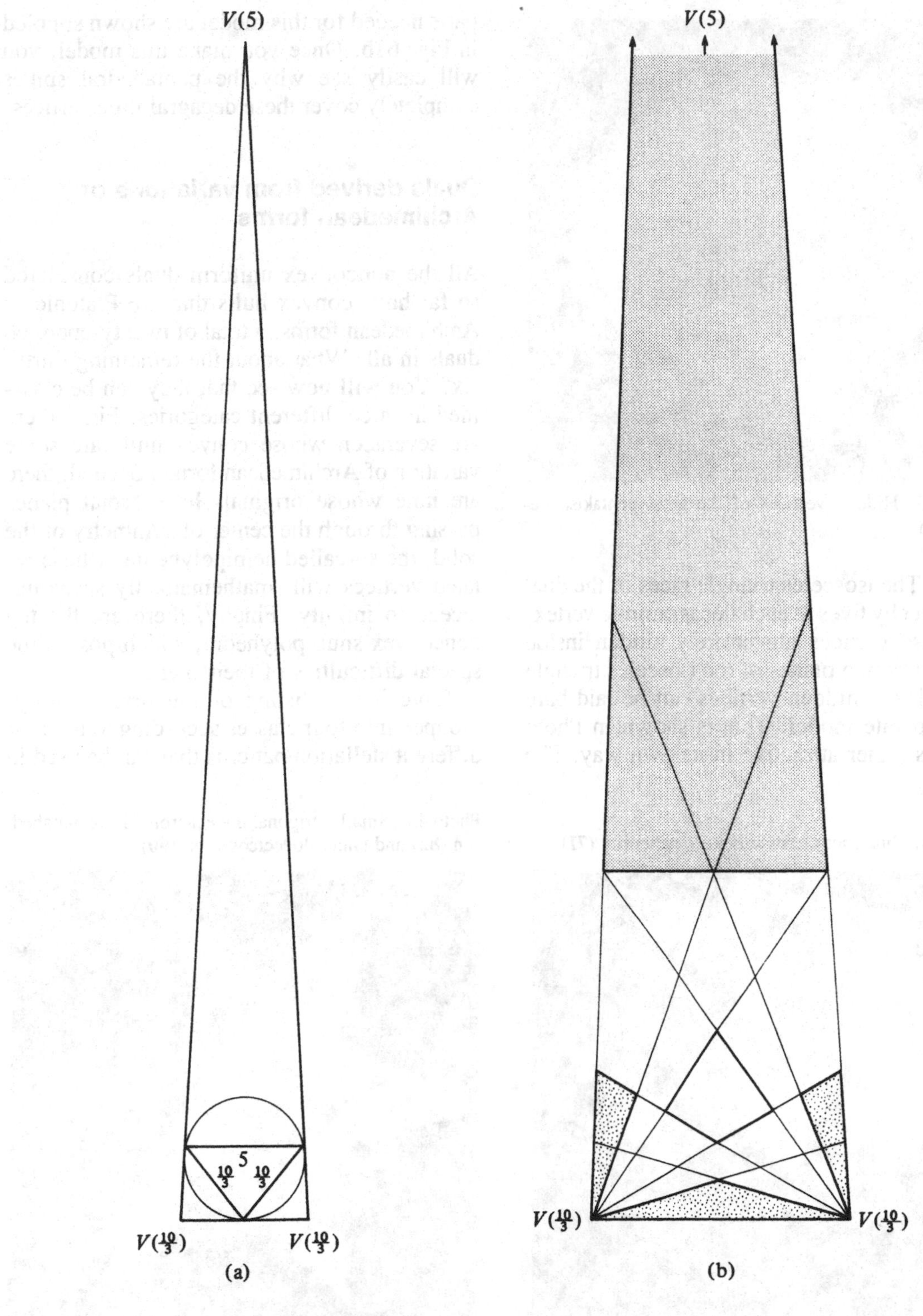

Fig. 61. Dorman Luke construction and stellation pattern for dual of **97**.

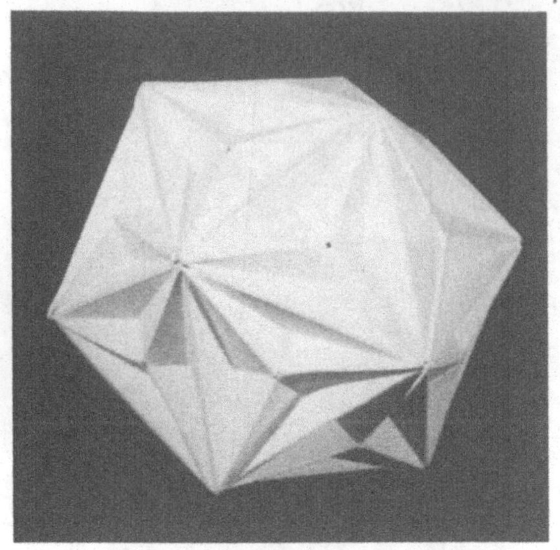

Photo 43. Hidden vertices of the great pentakisdodecahedron.

center. The isosceles triangle faces of the dual also meet by fives at each decagrammic vertex, but these vertices are entirely hidden inside the intersecting planes of the isosceles triangle faces. These hidden vertices can be laid bare in a separate model. This is shown in Photo 43. It is rather attractive in its own way. The parts needed for this model are shown stippled in Fig. 61b. Once you make this model, you will easily see why the pentahedral spikes completely cover these decagrammic vertices.

Duals derived from variations of Archimedean forms

All the nonconvex uniform duals considered so far have convex hulls that are Platonic or Archimedean forms, a total of twenty-one such duals in all. What about the remaining thirty-six? You will now see that they can be classified in three different categories. First, there are seventeen whose convex hulls are some variation of Archimedean forms. Second, there are nine whose originals have facial planes passing through the center of symmetry of the solid, the so-called hemipolyhedra, whose related vertices will, mathematically speaking, recede to infinity. Finally, there are the ten nonconvex snub polyhedra, which pose some special difficulties of their own.

Table 3 is a listing of the first category, grouped into four classes according to the four different stellation patterns that can be used in

Photo 44. Small icosacronic hexecontahedron (**71**).

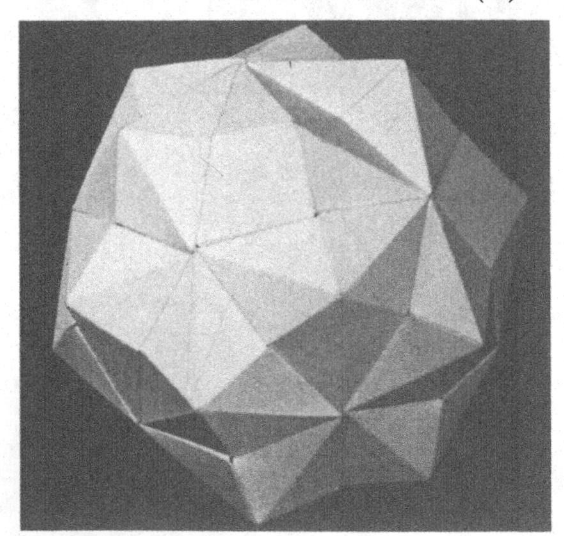

Photo 45. Small ditrigonal dodecacronic hexecontahedron (**82**) and small dodecicosacron (**90**).

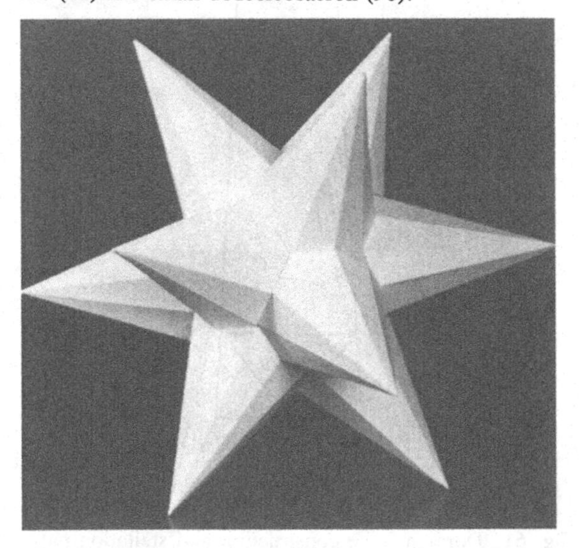

Table 3. *Duals derived from variations of Archimedean forms*

Polyhedron	Stellation pattern	Convex hull	Dual of hull	Dual
71	Fig. 62	Variations of the rhombicosidodeca	Deltoidal hexeconta	Photo 44
82				Photo 45
90				Photo 45
95				Photo 46
104				Photo 47
75	Fig. 70	Variations of the truncated icosa	Pentakis dodeca	Photo 48
76				Photo 49
83				Photo 50
96				Photo 51
99				Photo 52
105				Photo 53
109				Photo 53
79	Fig. 78	Variations of the rhombitruncated cubocta	Hexakis octa	Photo 54
93				Photo 55
84	Fig. 81	Variations of the rhombitruncated icosidodeca	Hexakis icosa	Photo 56
98				Photo 57
108				Photo 58

making models. Strictly speaking, there should actually be more stellation patterns than four, but for the sake of model making it will be best to introduce some compromises. For example, Fig. 62a shows a stellation pattern from which the duals of **71**, **82**, and **90** are derived, but the same pattern can be used for the duals of **95** and **104**. If you look back at Fig. 58 and compare it with Fig. 62a, you will see that it differs in no way except that the latter is a fuller or more extended stellation pattern than the former. The compromises introduced here are really not in the mathematics, but only in the drawings used for making the models. This will save a great deal of work, and actual calculations can still be done in each instance. You are invited to do these on your own and compare results.

An accurate stellation pattern for the duals of **71**, **82**, **90**, and **104** is shown in Fig. 62b. If you compare Photo 47a with Photo 47b, you will see the slight variation that has taken place in the dual of **104**.

Photo 46. Great stellapentakisdodecahedron (**95**).

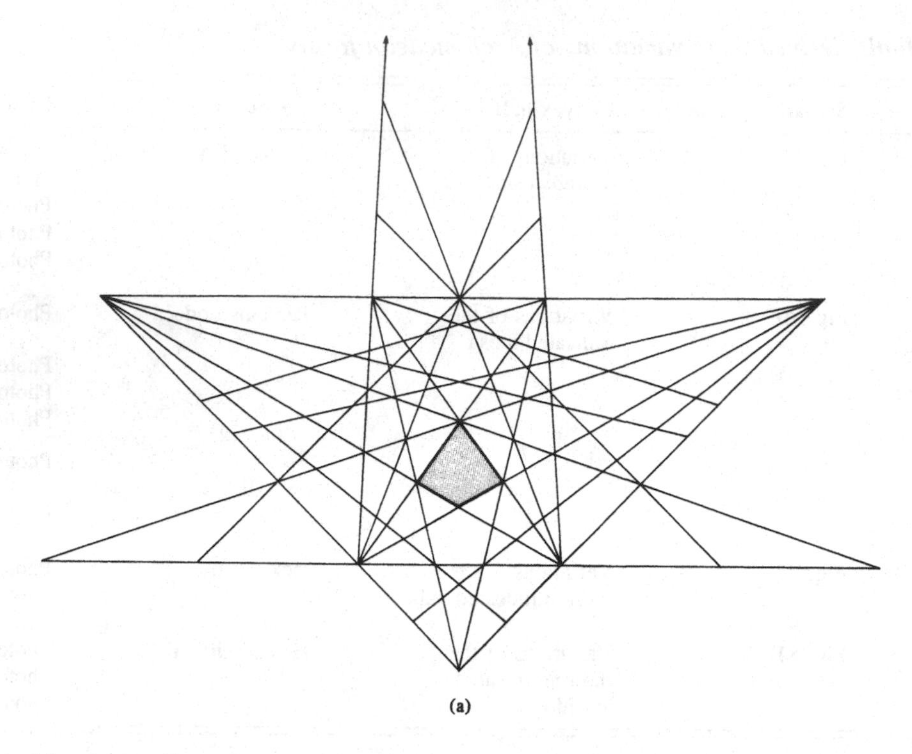

Fig. 62a. Archimedean stellation pattern for approximate duals of **71**, **82**, **90**, **95**, and **104**.

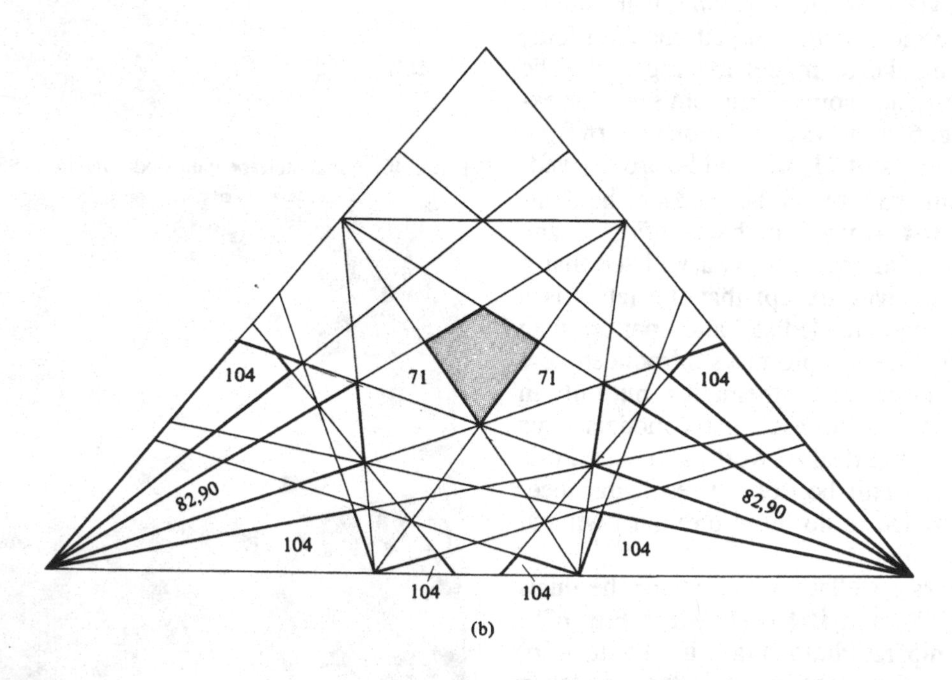

Fig. 62b. Stellation pattern for exact duals of **71**, **82**, **90**, and **104**.

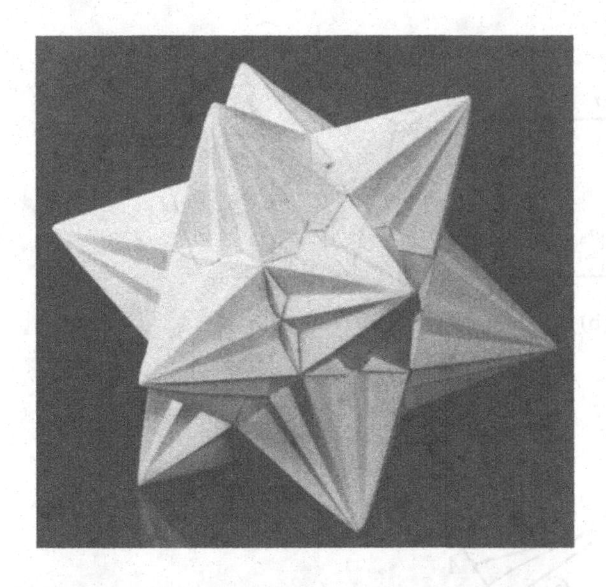

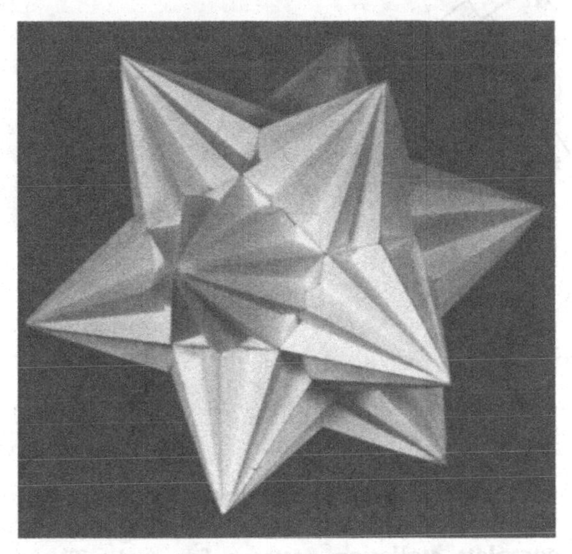

Photo 47a,b. Great triakisicosahedron (**104**).

You are already aware that Fig. 58 is a partial stellation pattern from which the duals of **72**, and **74**, and **97** were derived. The convex core of this stellation is the Archimedean deltoidal hexecontahedron, dual of the Archimedean rhombicosidodecahedron. But now look carefully at the original uniform polyhedron **71**. Its convex hull is also a rhombicosidodecahedron, where the triangles form part of this hull but where the pentagrams must be covered by pentagons and where the intervening grooves become rectangles, not squares. So this is a variation of the Archimedean rhombicosidodecahedron. Interestingly enough, dihedral angles between faces remain constant, and full symmetry is maintained. Only measured lengths of edges change in relation to each other. What, then, you may be asking, is the dual of this convex hull? This is where the polar reciprocal formula comes into its own and achieves its full usefulness. You are invited to work out for yourself the full mathematical accuracy called for in this instance by studying Fig. 63. Three faces of the hull for **71** are shown in Fig. 63a–c, each with applicable edge lengths given, from which the measures of apothems for these faces may be calculated. Next, a layout is shown in Fig. 63d, with a typical vertex shown in Fig. 63e to help visualize how this layout receives its data so that the polar reciprocal formula can be applied. The measure of the radius of the layout, ${}_0R$, comes from the proportionality relationship

$$\frac{{}_0R}{\tau} = \frac{[(17 + 3\sqrt{5})/2]^{1/2}}{2}$$

See *Polyhedron models* for numerical data relative to edge length and radius. Finally, using the appropriate trigonometric formulas, you can verify the fact that the face of the dual of this convex hull must have the shape shown in Fig. 63f. The angles of this deltoidal face turn out to be 69.353522, 86.825041 (taken twice), and 116.996396. Comparing these results with the Archimedean dual, which has face angles of 67.78301, 86.974155 (taken twice), and 118.26868, you see that they are so nearly the same shape that you would be hard put to draw them accurately enough to be able to see the difference, let alone the difference in the convex dual models if they were made and set side by side for comparison. Hence, for purposes of model making, practically nothing is lost by using the same stellation pattern for the duals of **71**, **82**, and **90**, as well as for **72**, **74**, and **97**.

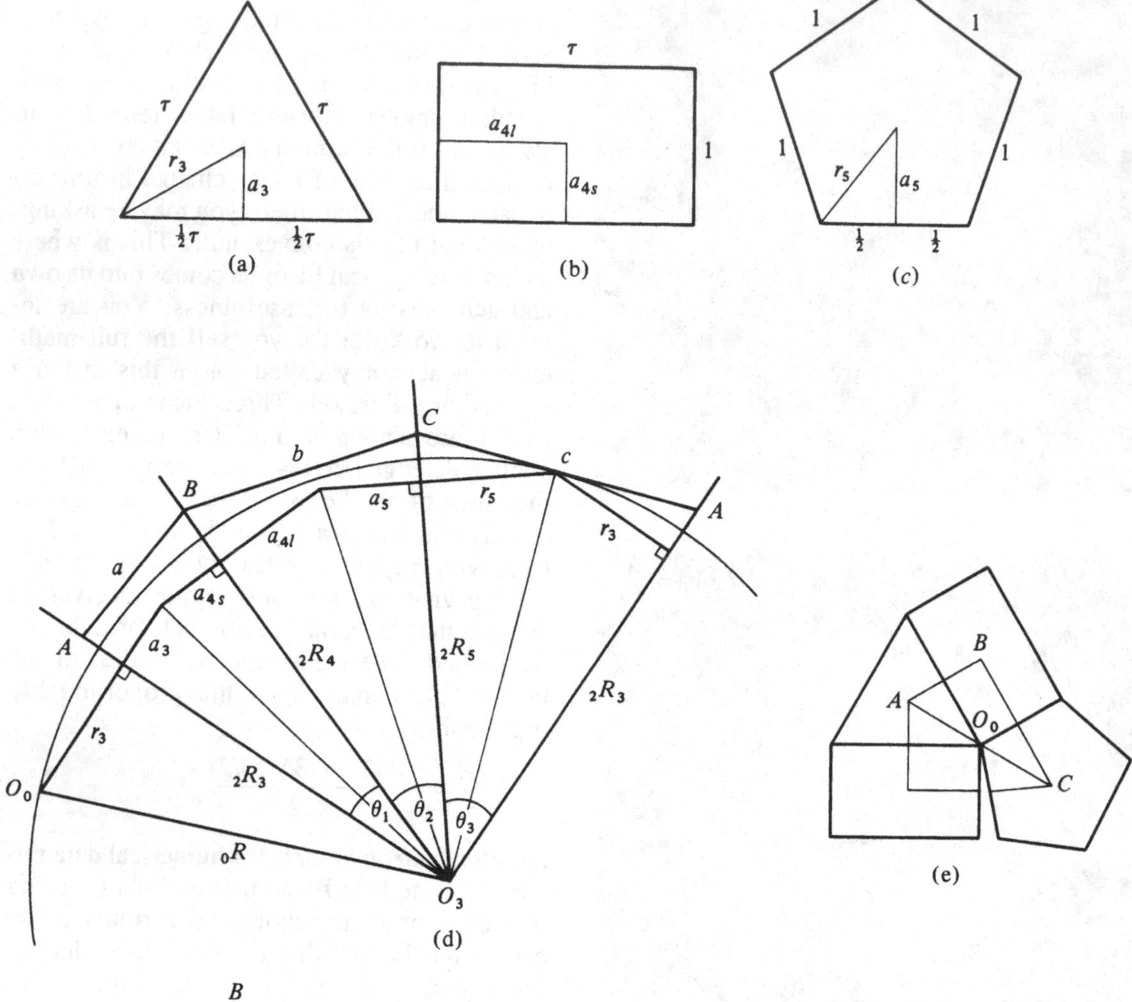

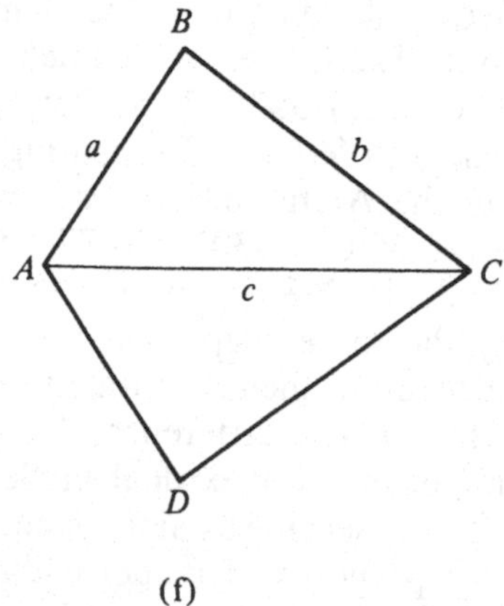

Fig. 63. Polar reciprocation applied to **71**.

A second advantage of doing models in this way is that you can avoid working out the complete stellation patterns for each. These patterns would show only slight variations that might even be imagined as the consequences of a continuous transformation of the Archimedean form into other shapes, retaining, however, dihedral angles and symmetry axes constant. Accurate stellation patterns are very difficult to produce without numerical data, which can come only from the solution of algebraic equations that define the facial planes involved. This is where computer analysis can yield great savings in time and labor if mathematical accuracy is the primary goal.

Table 3 shows other sets of triplets, related because of their vertex figures and convex hulls. Photos 45 and 53 are again an instance where one model is the dual of two different originals because of the placement of hidden vertices. Furthermore, careful examination of the uniform polyhedron **104** will show that its convex hull actually has the same shape as the convex hulls of **71**, **82**, and **90**. But **95** is again a slightly different variation. You have plenty of challenging work to do here if you want to verify these details for yourself.

Figures 64 through 68 are the drawings for the Dorman Luke construction and the facial planes for each dual listed at the beginning of Table 3. The shaded parts show the exterior portions needed for making a model. If these models are not absolutely accurate, they will at least be topologically very similar to any models more accurately made from mathematically calculated data and drawings.

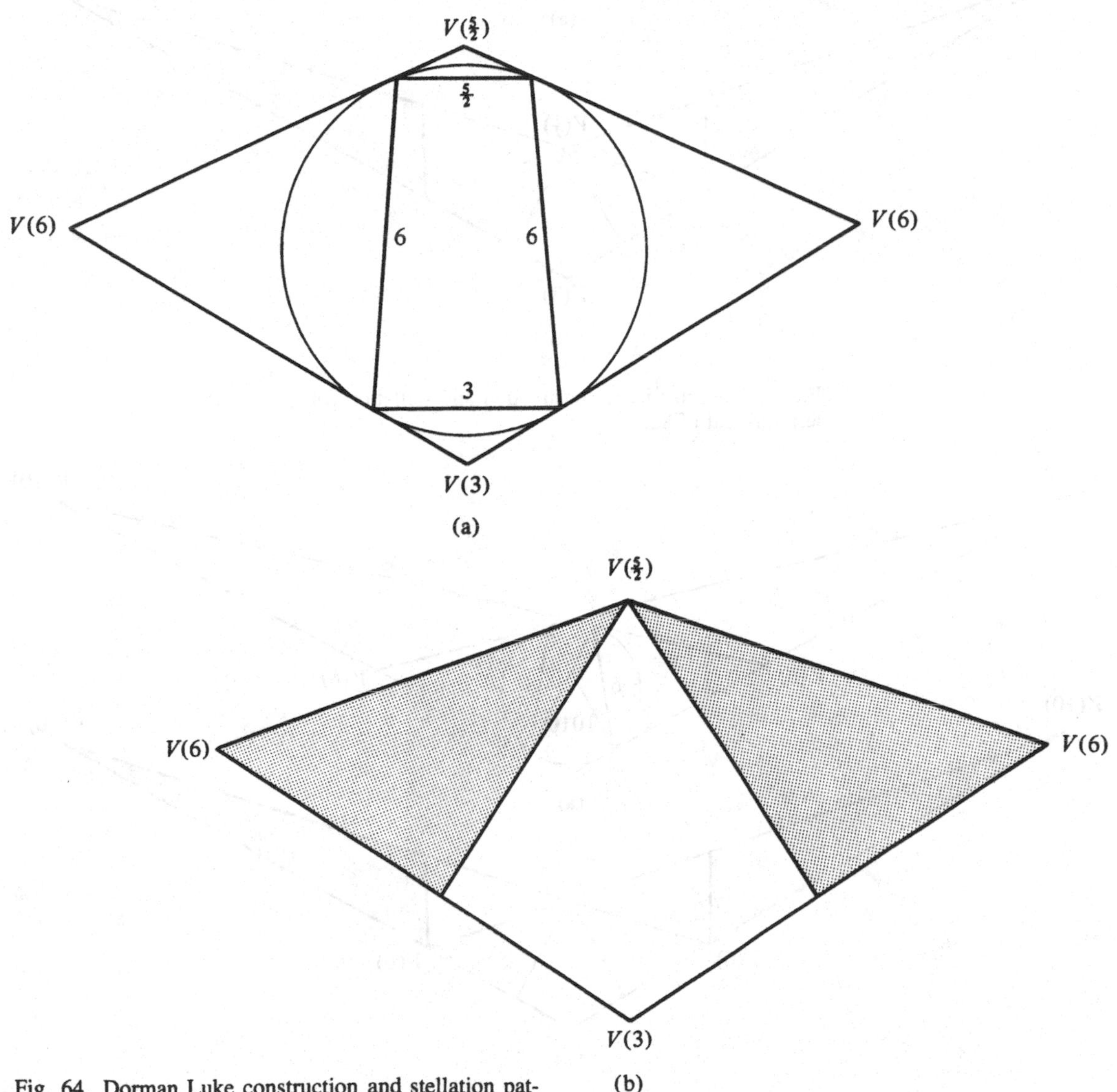

Fig. 64. Dorman Luke construction and stellation pattern for dual of **71**.

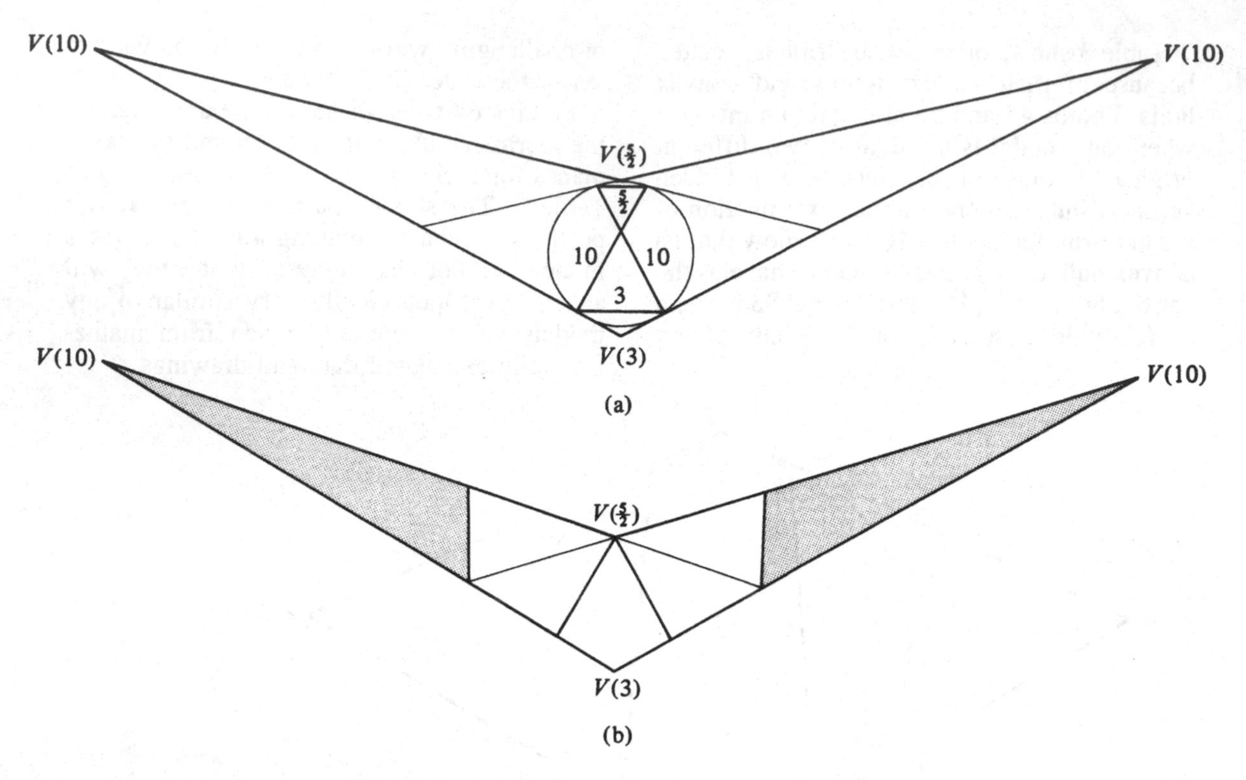

Fig. 65. Dorman Luke construction and stellation pattern for dual of **82**.

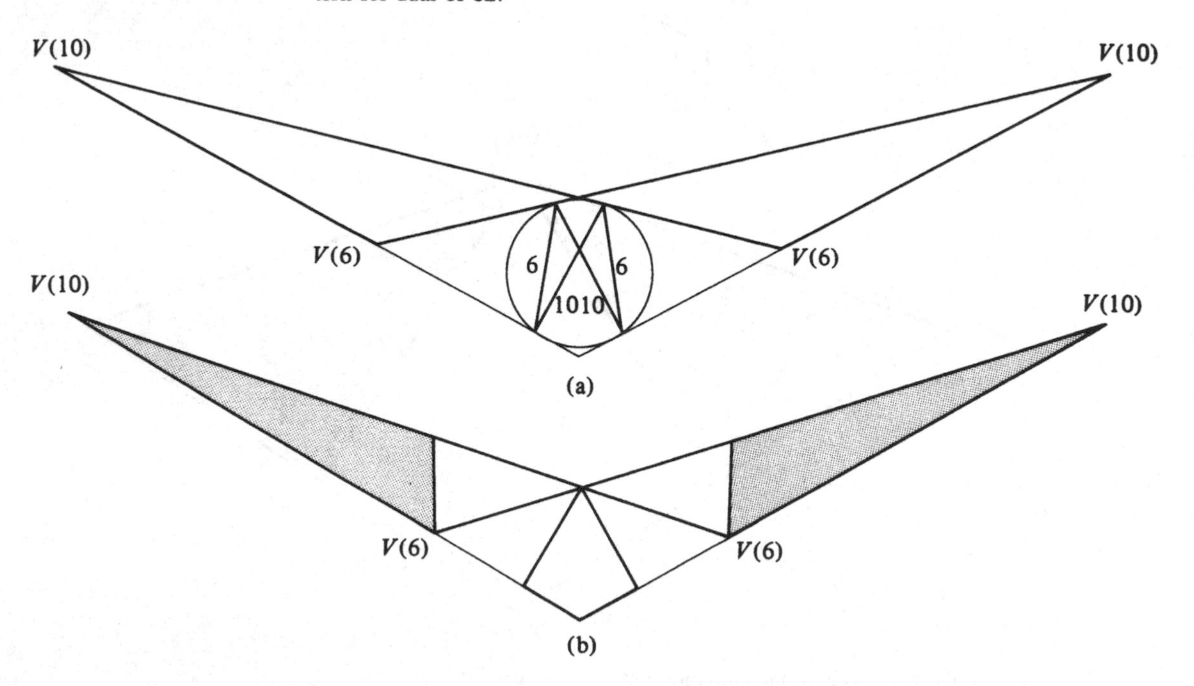

Fig. 66. Dorman Luke construction and stellation pattern for dual of **90**.

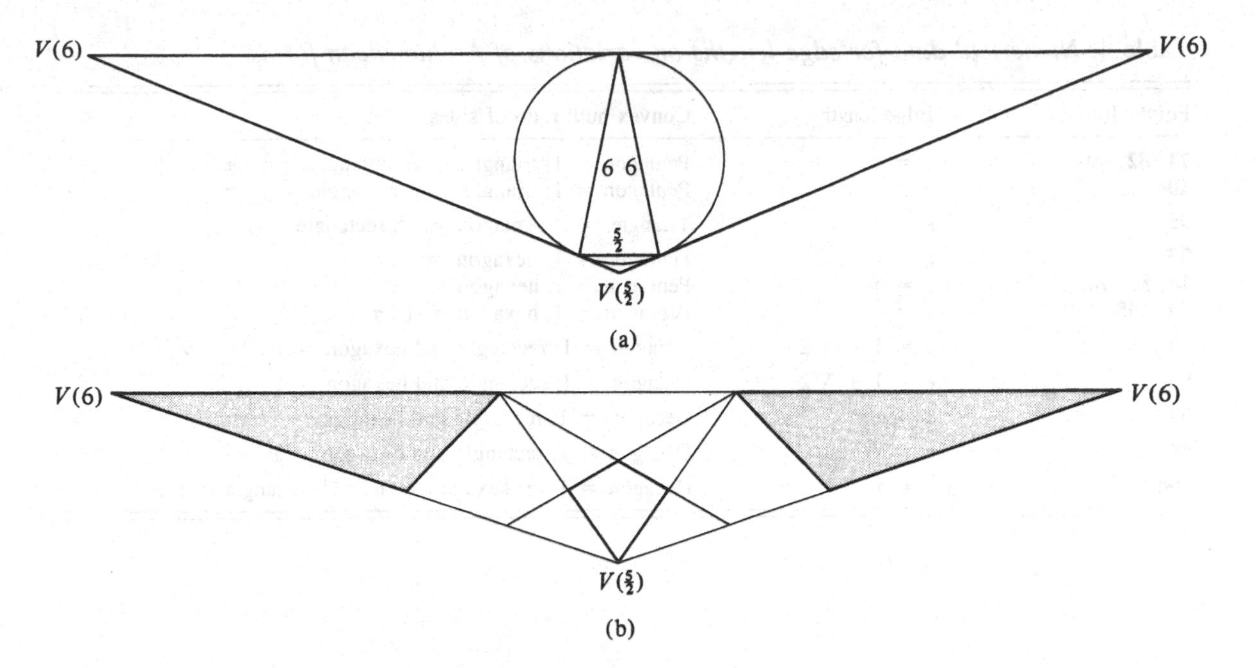

Fig. 67. Dorman Luke construction and stellation pattern for dual of **95**.

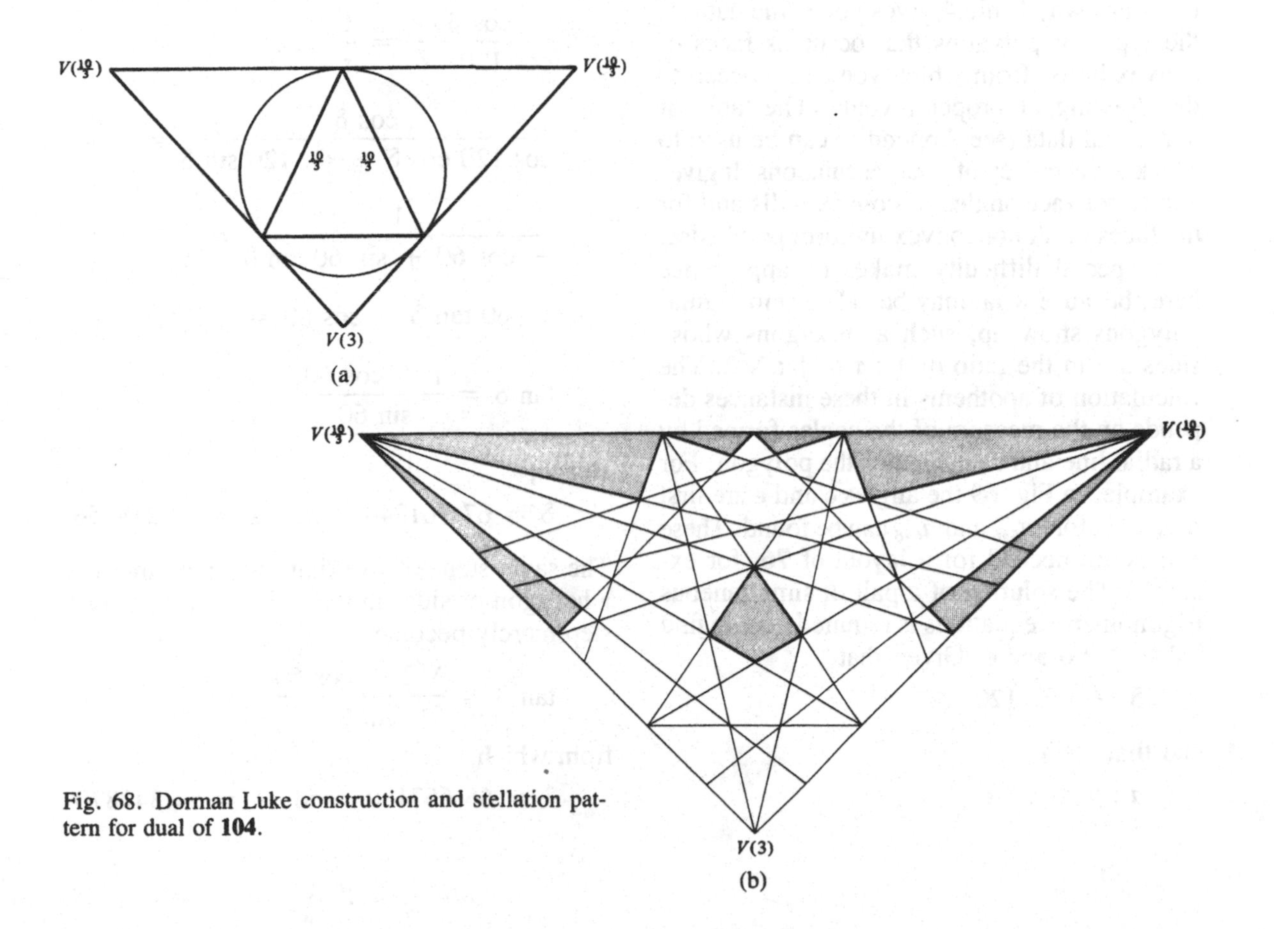

Fig. 68. Dorman Luke construction and stellation pattern for dual of **104**.

Table 4. *Numerical data for edge lengths on variations of Archimedean forms*

Polyhedron	Edge length	Convex hull: ratio of sides
71, 82, 90	$e = \tau$	Pentagon = 1; triangle = τ; rectangle = $1 : \tau$
104	$e = \tau^3$	Pentagon = 1; triangle = τ; rectangle = $1 : \tau$
95	$e = \tau^3$	Triangle = 1; pentagon = τ^2; rectangle = $1 : \tau^2$
75	$e = \tau$	Pentagon = 1; hexagon = $1 : \tau$
76, 83, 96	$e = \tau^2$	Pentagon = τ; hexagon = $1 : \tau$
99, 105, 109	$e = \tau^3$	Pentagon = 1; hexagon = $1 : \tau$
79	$e = 1 + \sqrt{2}$	Octagon = 1; rectangle and hexagon = $1 : 1 + \sqrt{2}$
93	$e = 1 + \sqrt{2}$	Octagon = 1; rectangle and hexagon = $1 : \sqrt{2}$
84	$e = \tau^2$	Decagon = 1; rectangle and hexagon = $1 : \tau^2$
98	$e = \tau^2$	Decagon = 1; rectangle and hexagon = $1 : \tau$
108	$e = \tau^3$	Decagon = $1 : \tau$; hexagon = $1 : \tau^{-1}$; rectangle = $\tau : \tau^{-1}$

If you want to do some of these calculations on your own, Table 4 gives you some data on the types of polygons that occur as faces of convex hulls, from which you can proceed to the drawing of proper layouts. The table of numerical data (see Appendix) can be used to check the accuracy of your calculations. It gives values for face angles of convex hulls and for the faces of all nonconvex uniform polyhedra.

A special difficulty makes its appearance here, because what may be called semiregular polygons show up, such as hexagons whose sides are in the ratio of $1 : \tau$ or $1 : \sqrt{2}$. The calculation of apothems in these instances depends on the measure of the angles formed by a radial line and each side of the polygon. For example, in Fig. 69 the angles δ and ϵ are first needed before $a_{5,6}$ and $a_{6,6}$ can be found. These values are needed for a layout of **76**, for example. The solution of a pair of simultaneous trigonometric equations is required here to find values for δ and ϵ. Given that

$$\delta + \epsilon = 120$$

and that

$$a : b = 1 : \tau$$

it follows that

$$\frac{\cos \delta}{\cos(120-\delta)} = \frac{1}{\tau}$$

$$\frac{\cos \delta}{\cos 120 \cos \delta + \sin 120 \sin \delta} = \frac{1}{\tau}$$

$$\frac{1}{-\cos 60 + \sin 60 \tan \delta} = \frac{1}{\tau}$$

$$\sin 60 \tan \delta - \cos 60 = \tau$$

$$\tan \delta = \frac{\tau + \cos 60}{\sin 60}$$

Hence,

$$\delta = 67.761244 \quad \text{and} \quad \epsilon = 52.238756$$

The same steps of working will show that for a hexagon of sides in the ratio $1 : \sqrt{2}$ the last step merely becomes

$$\tan \delta = \frac{\sqrt{2} + \cos 60}{\sin 60}$$

from which

$$\delta = 65.657130 \quad \text{and} \quad \epsilon = 54.342870$$

You can now see how this can be generalized for other types of hexagons; namely,

$$\tan \delta = \frac{k + \cos 60}{\sin 60}$$

where k may take on any of these values: τ; $\sqrt{2}$; $1 + \sqrt{2}$; $1 + \tau$; τ^{-1}.

It is interesting to notice that for the octahedral symmetry group to which **79** and **93** belong it is the number $\sqrt{2}$ that plays a predominant role, and in the icosahedral symmetry group it is the number τ. Notice, too, that the triplet of **76**, **83**, and **96**, along with **75**, displays the golden-section truncation of the icosahedron, some of whose stellations were shown in Photos 29 and 30. Here it is the dual of this golden-section truncation that becomes the core to be stellated in order to arrive at the mathematically correct duals of this set of polyhedra. But the other triplet, **99**, **105**, and **109**, will lose nothing from a practical point of view if you generate their duals from the same stellation pattern.

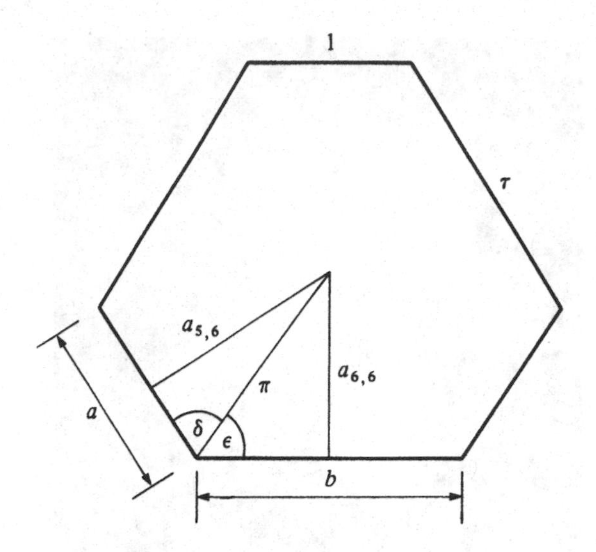

Fig. 69. Semiregular hexagon needed for calculation of apothems.

Figure 70 shows a stellation pattern from which the duals of the second group of polyhedrons listed in Table 3 can be derived. Models are shown in Photos 48 through 53.

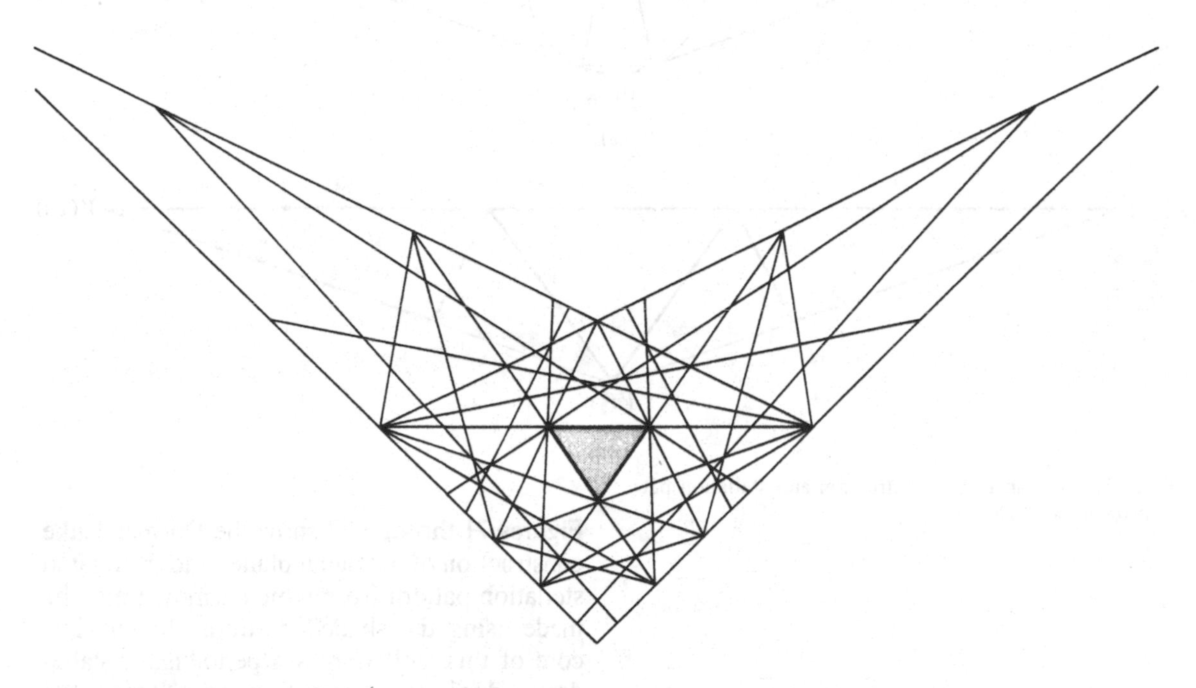

Fig. 70. Stellation pattern for duals of **75**, **76**, **83**, **96**, **99**, **105**, and **109**.

Photo 48. Small stellapentakisdodecahedron (**75**).

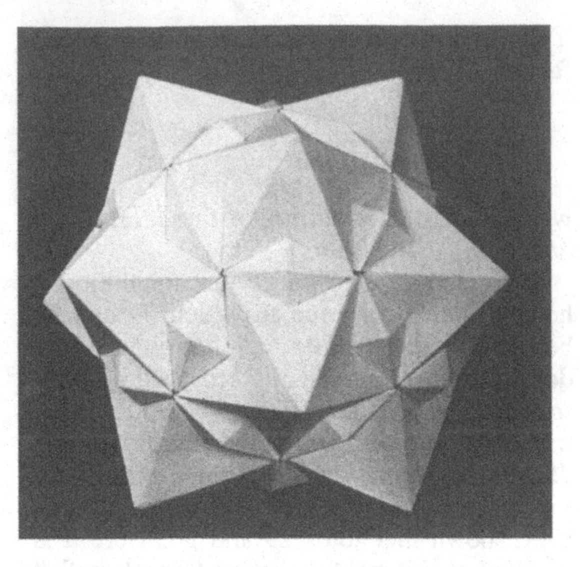
Photo 49. Medial deltoidal hexecontahedron (**76**).

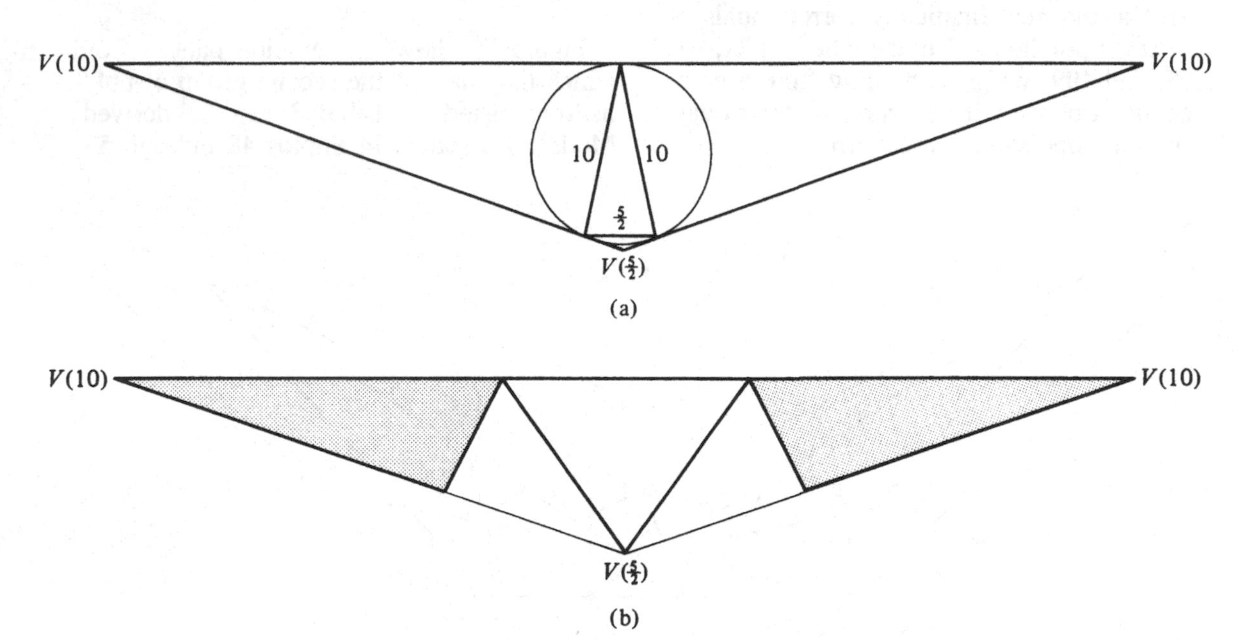

Fig. 71. Dorman Luke construction and stellation pattern for dual of **75**.

Figures 71 through 77 show the Dorman Luke construction of the facial planes and the related stellation pattern from which a model may be made using the shaded portions shown. The core of this stellation is a pentakisdodecahedron, dual of a variation of a truncated icosahedron.

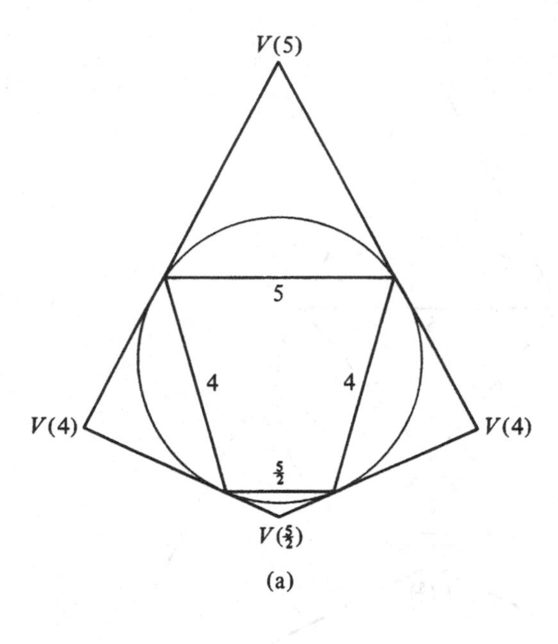

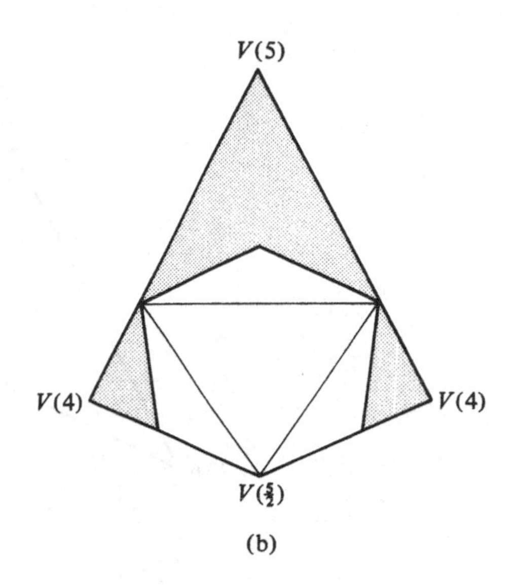

Fig. 72. Dorman Luke construction and stellation pattern for dual of **76**.

Photo 50. Medial icosacronic hexecontahedron (**83**).

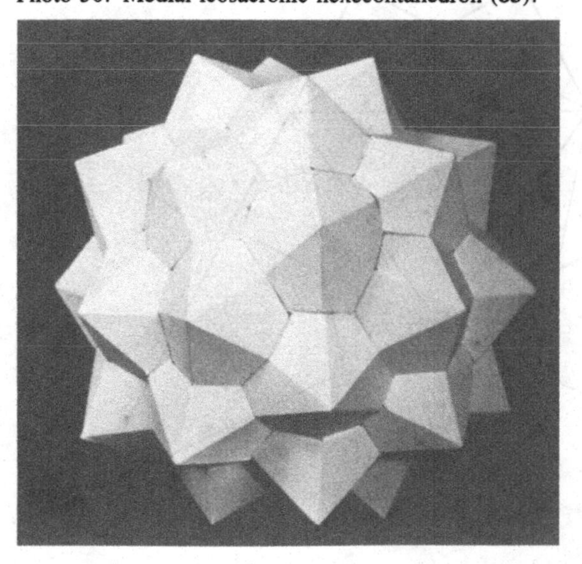

Photo 51. Rhombicosacron (**96**).

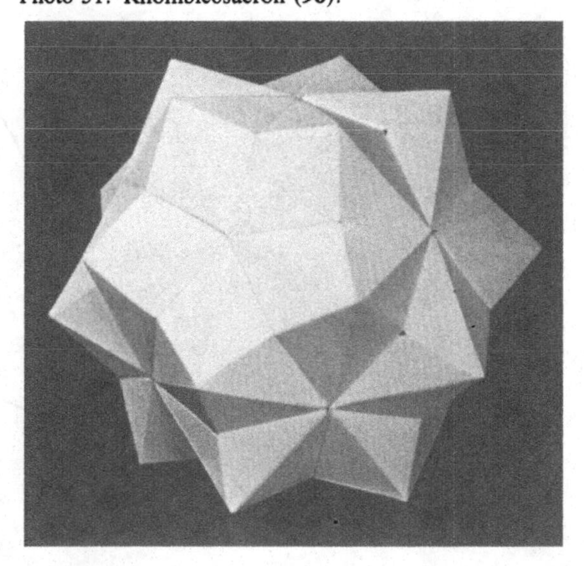

Fig. 73. Dorman Luke construction and stellation pattern for dual of **83**.

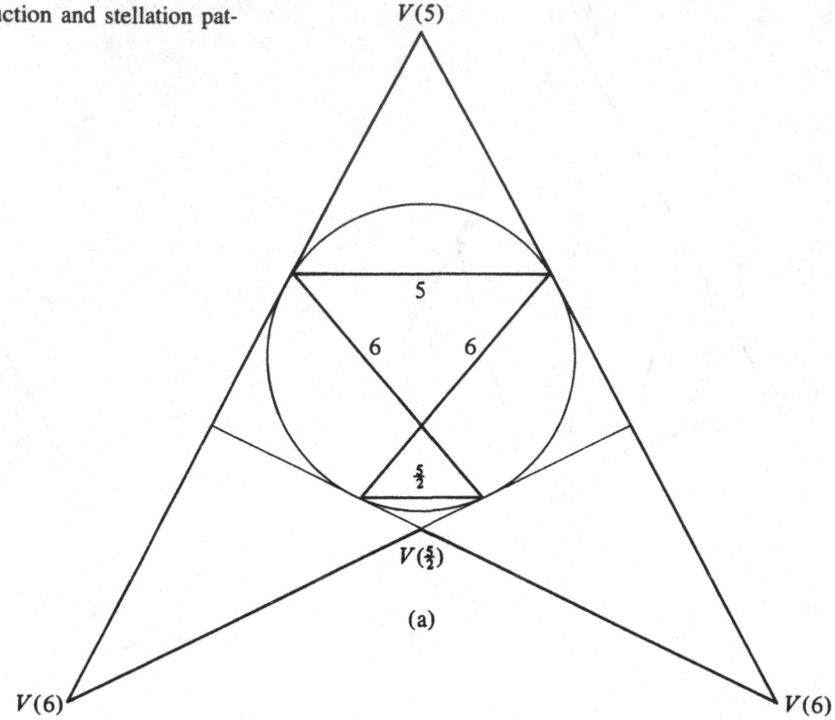

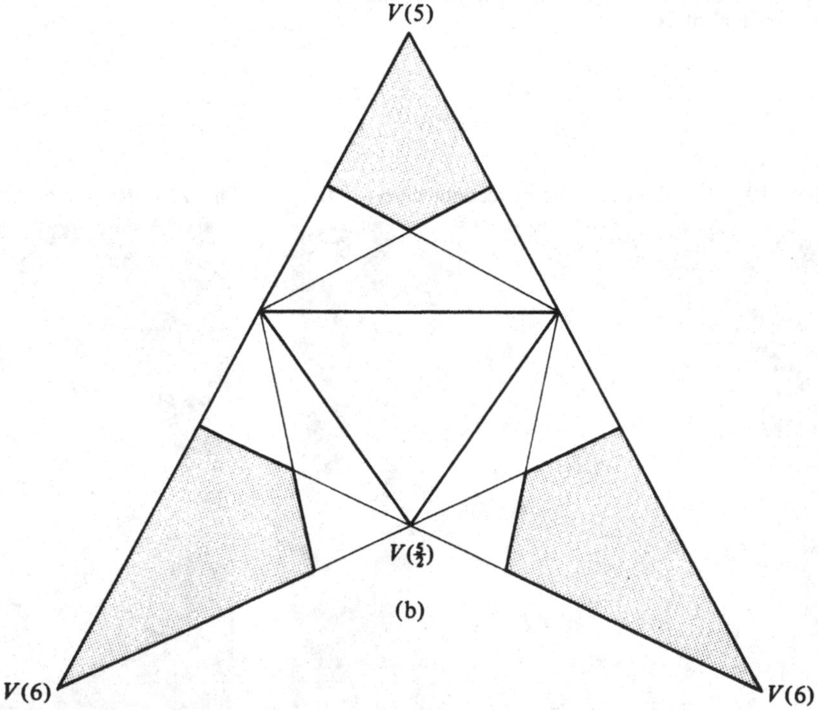

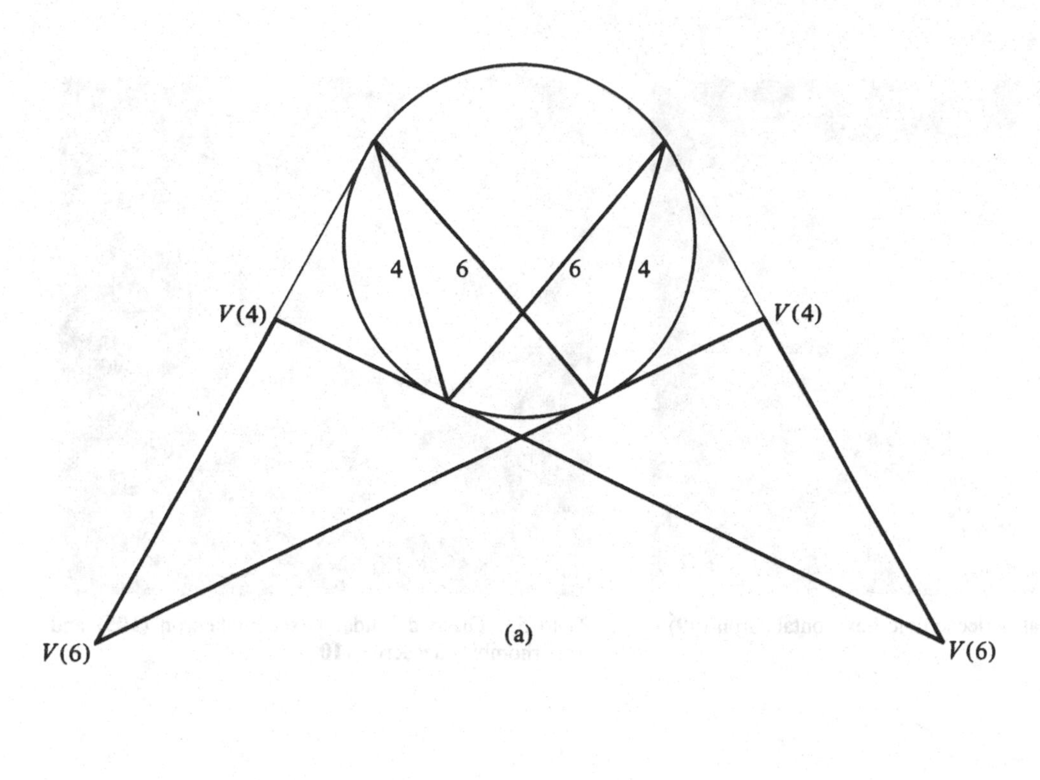

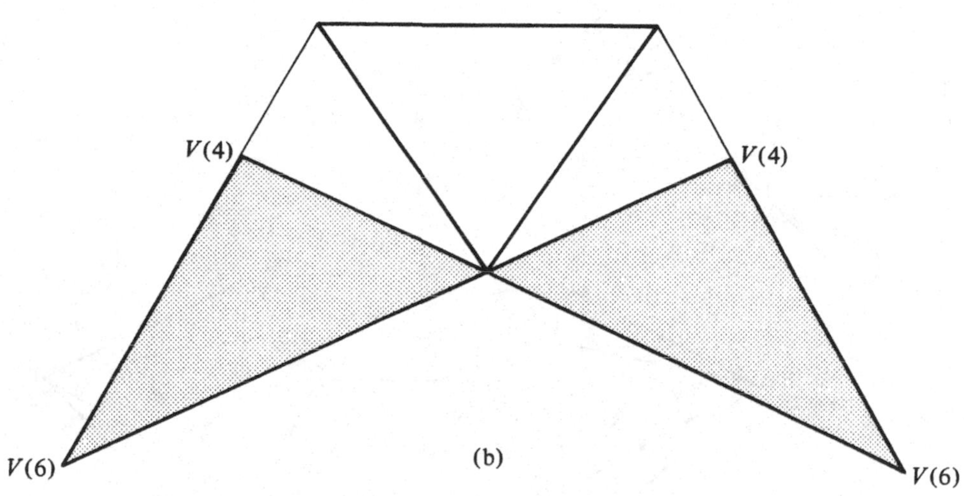

Fig. 74. Dorman Luke construction and stellation pattern for dual of **96**.

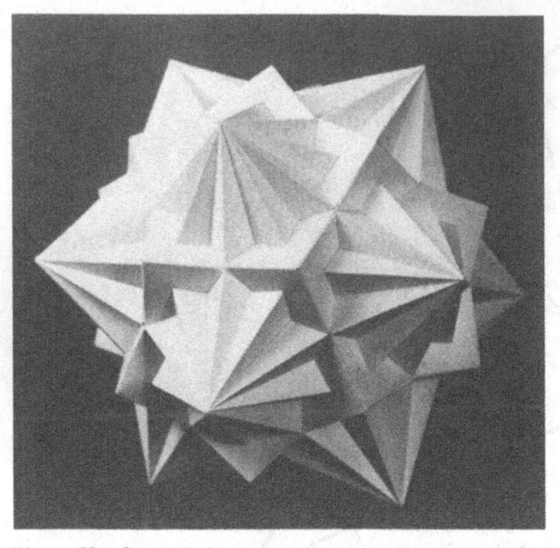

Photo 52. Great dodecacronic hexecontahedron (**99**).

Photo 53. Great deltoidal hexecontahedron (**105**) and great rhombidodecacron (**109**).

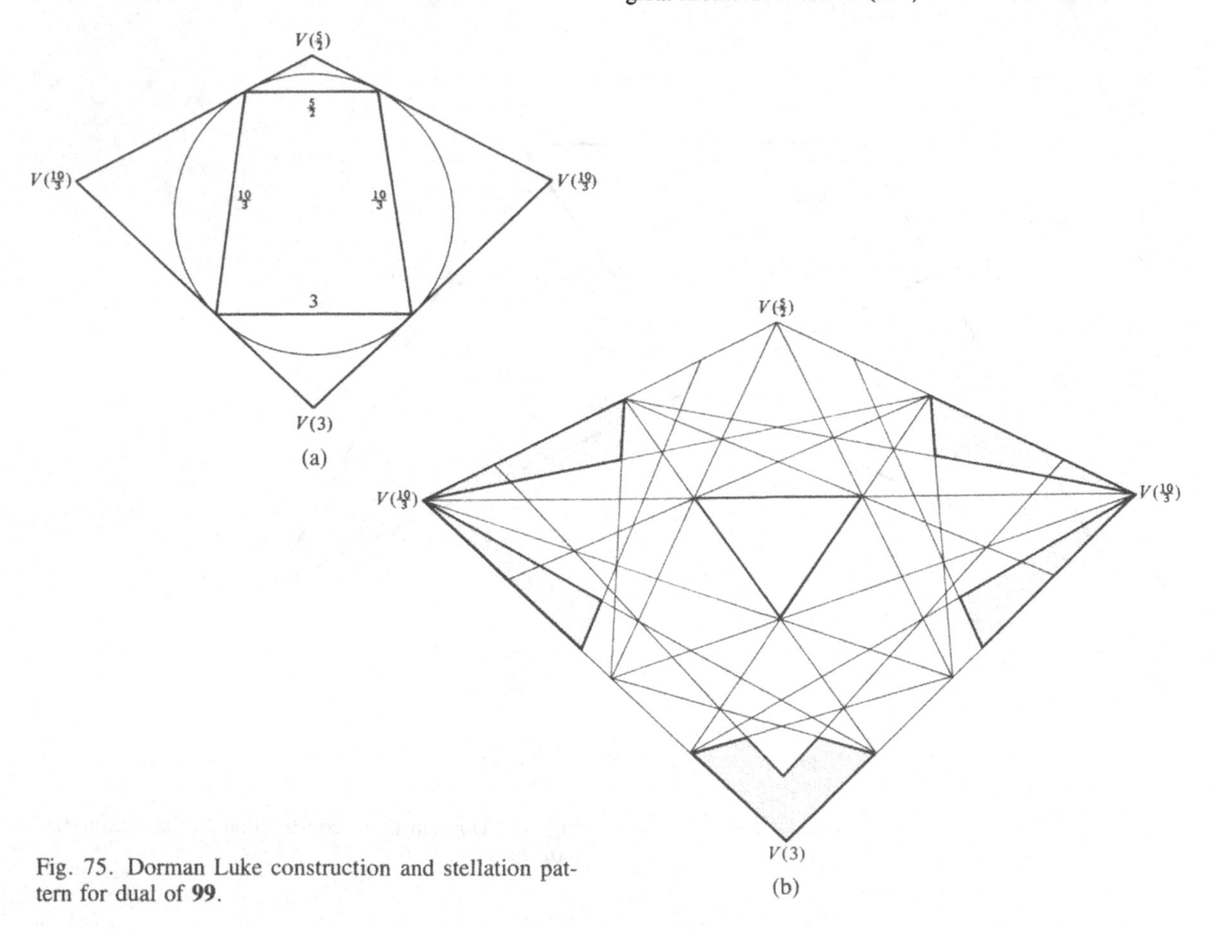

Fig. 75. Dorman Luke construction and stellation pattern for dual of **99**.

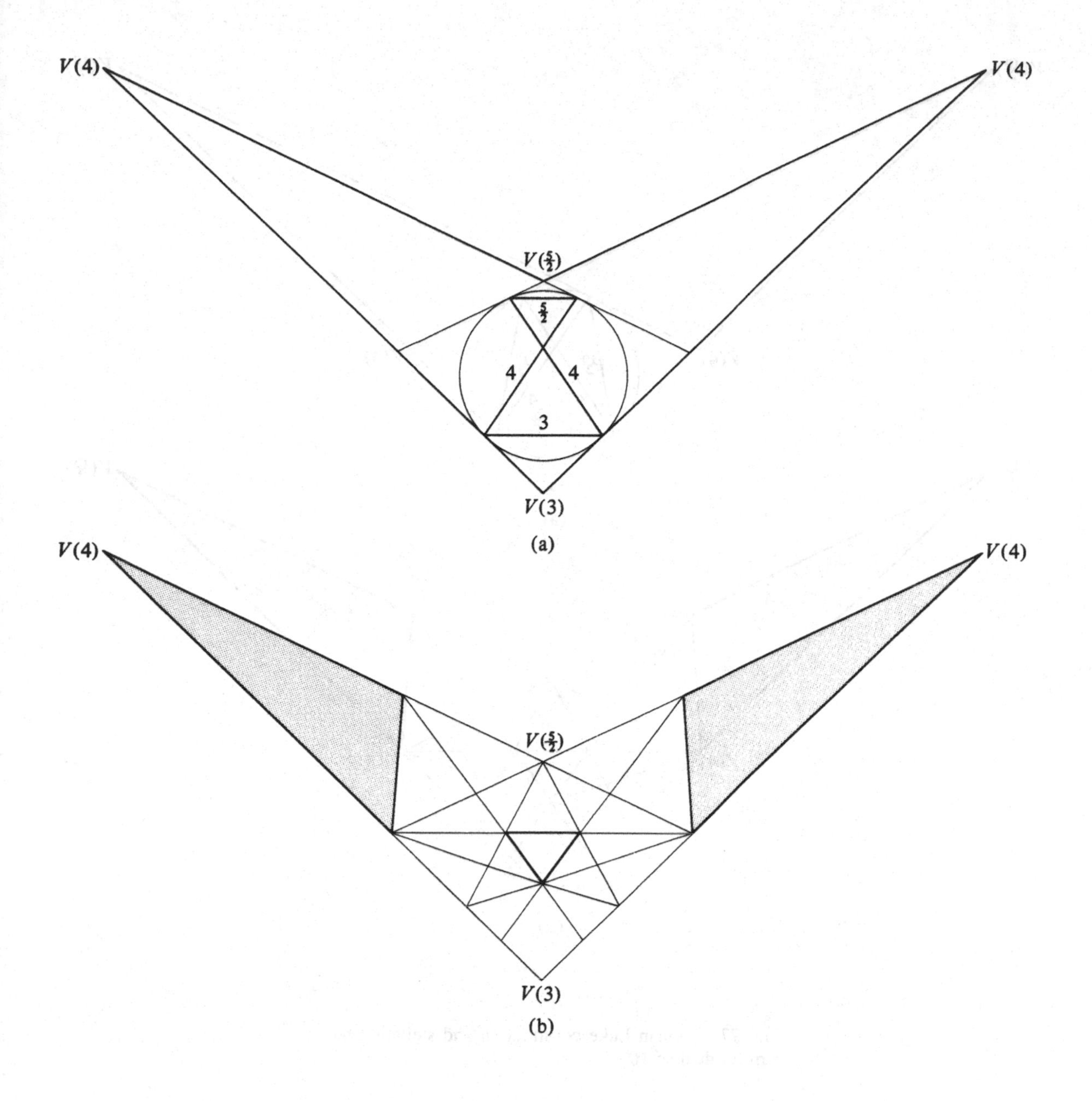

Fig. 76. Dorman Luke construction and stellation pattern for dual of **105**.

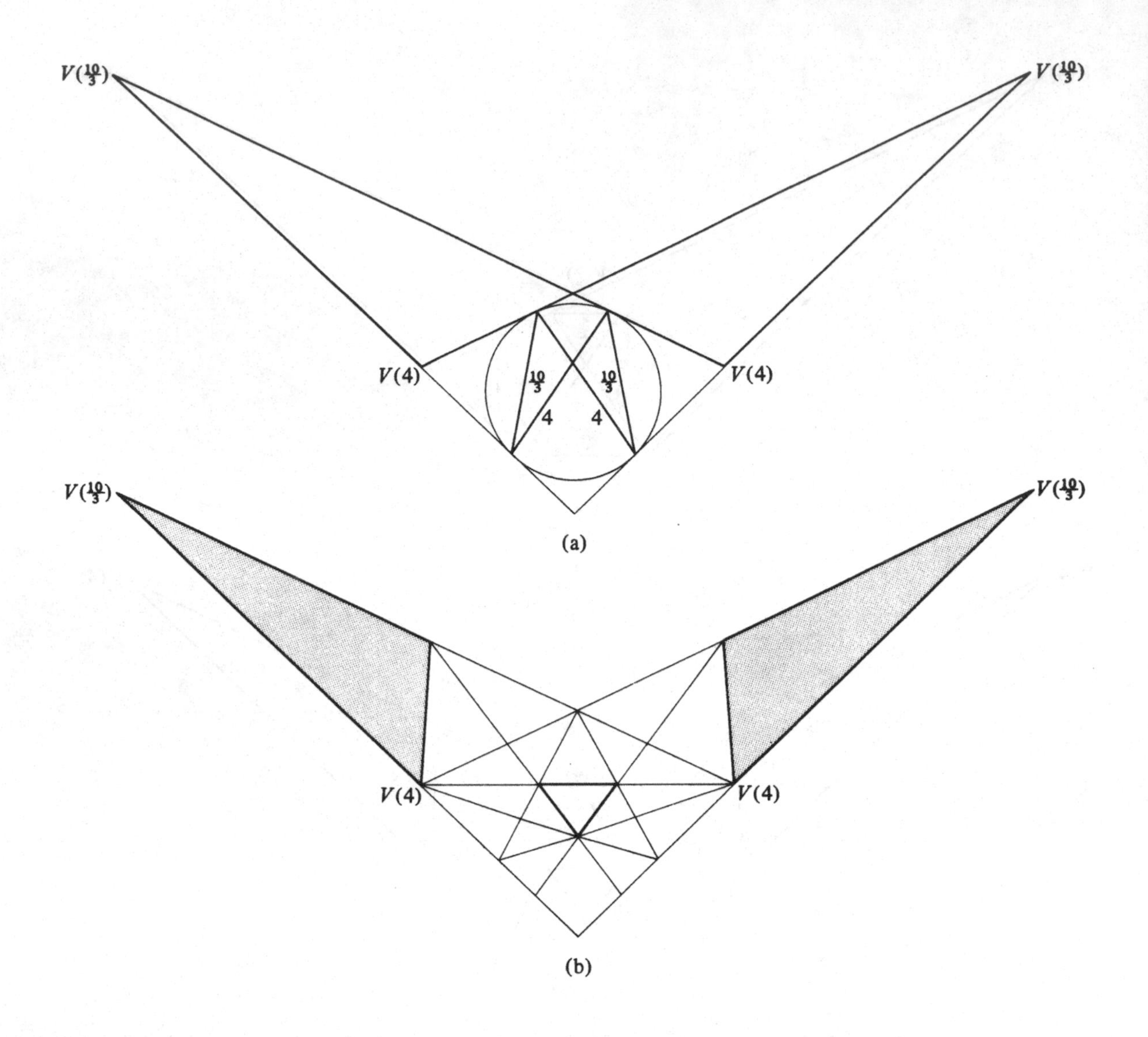

Fig. 77. Dorman Luke construction and stellation pattern for dual of **109**.

The uniform polyhedra **79** and **93** have convex hulls that are variations of the Archimedean rhombitruncated cuboctahedron, and the last three in Table 3, **84**, **98**, and **108**, have convex hulls that are variations of the Archimedean rhombitruncated icosidodecahedron. Once again, these differ mathematically from each other, so that it will actually be necessary to consider two different stellation patterns of the octahedral symmetry group and three different stellation patterns of the icosahedral symmetry group. For the sake of model making, the usual compromises are again introduced. Figure 78 shows a partial stellation pattern from which duals for **79** and **93** are generated. Figures 79 and 80 are the drawings from which the models shown in Photos 54 and 55 can be made. The model shown in Photo 54b is a truncated version of the one shown in Photo 54a. Figure 81 shows a partial stellation pattern for duals of **84**, **98**, and **108**, followed by Figs. 82, 83, and 84 for facial planes. Models are shown in Photos 56, 57, and 58.

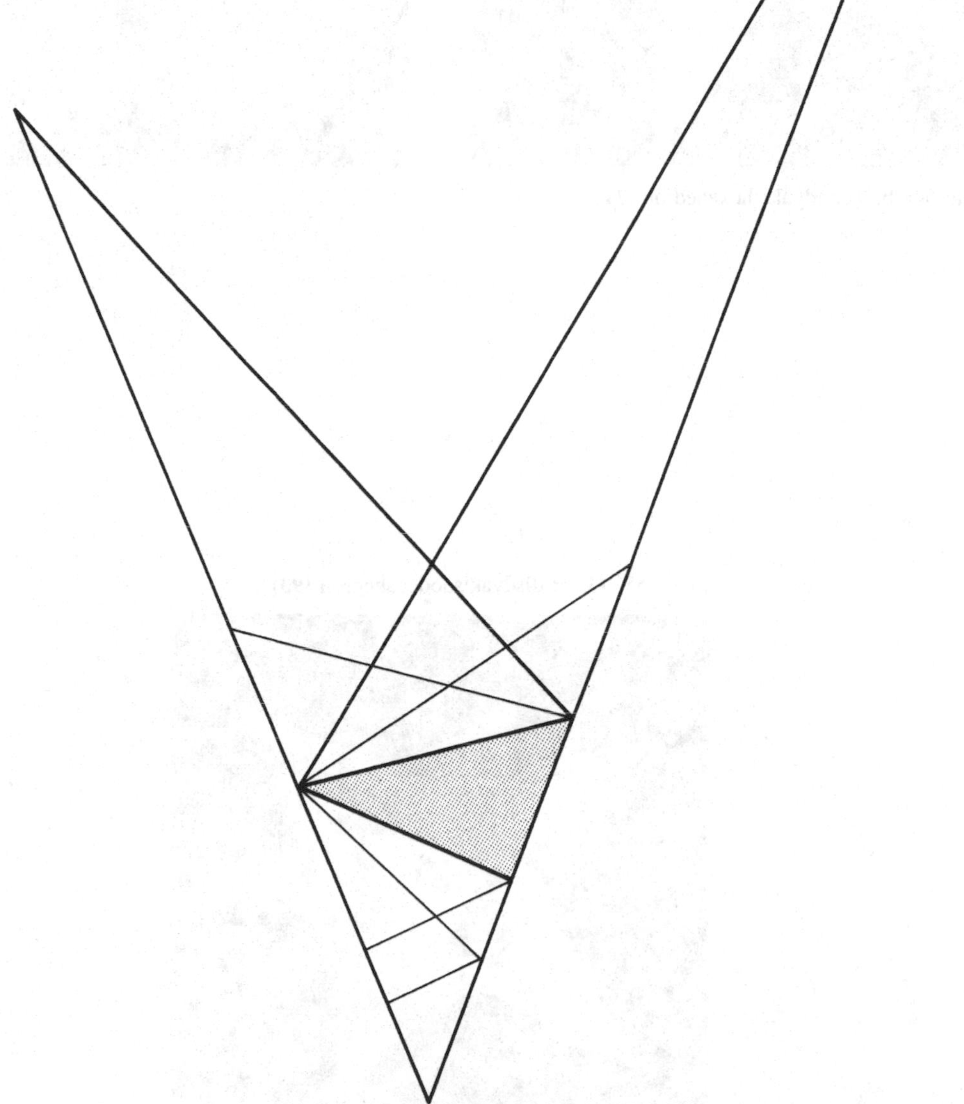

Fig. 78. Stellation pattern for duals of **79** and **93**.

Photo 54a,b. Tetradyakishexahedron (**79**).

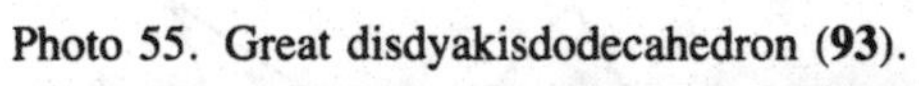

Photo 55. Great disdyakisdodecahedron (**93**).

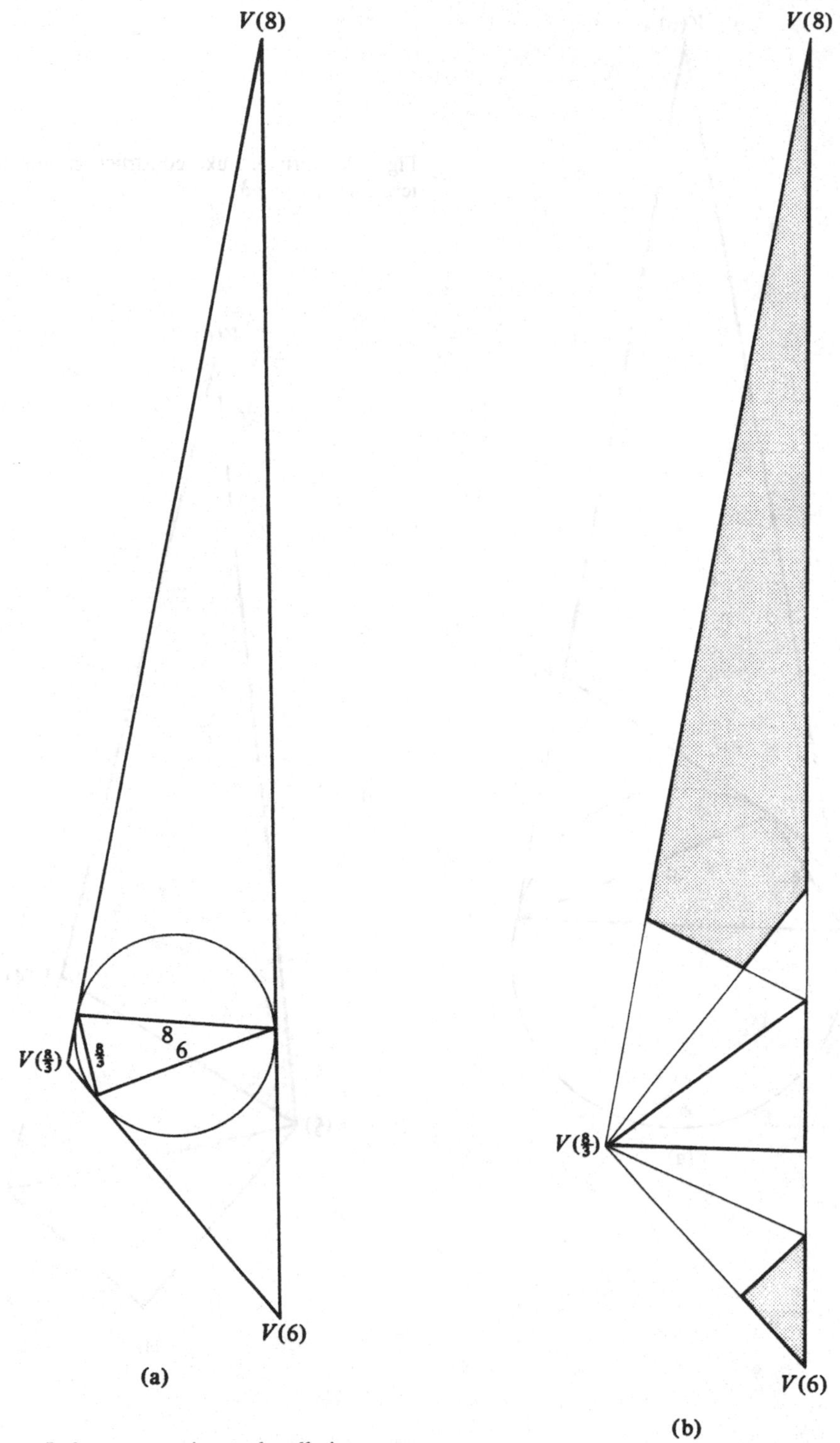

Fig. 79. Dorman Luke construction and stellation pattern for dual of **79**.

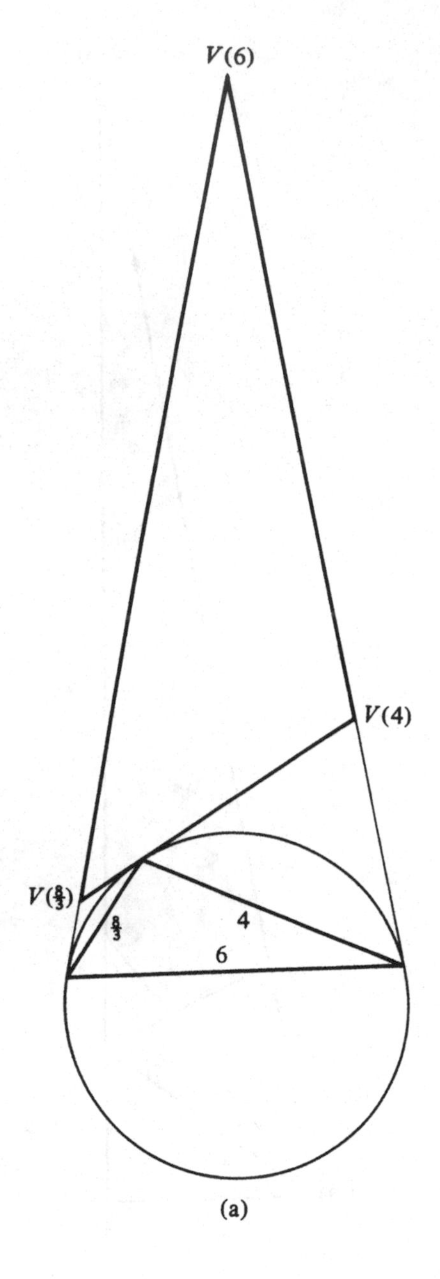

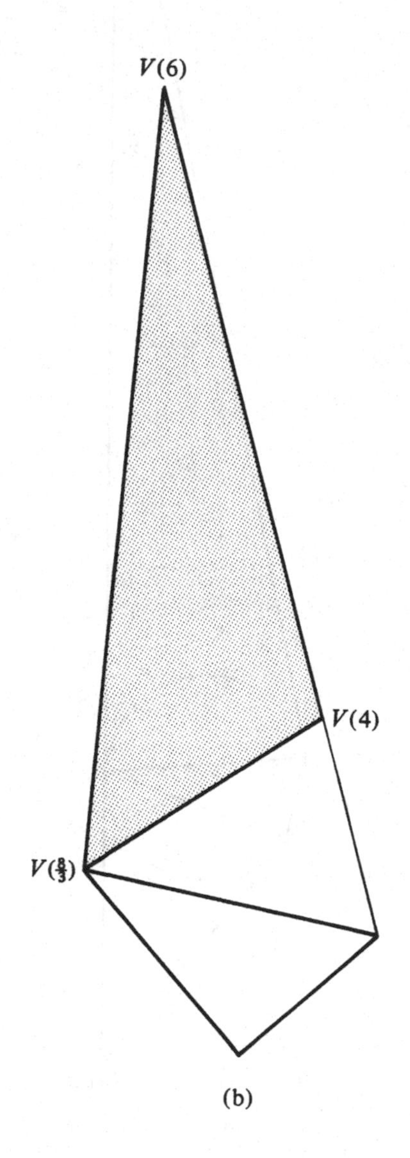

Fig. 80. Dorman Luke construction and stellation pattern for dual of **93**.

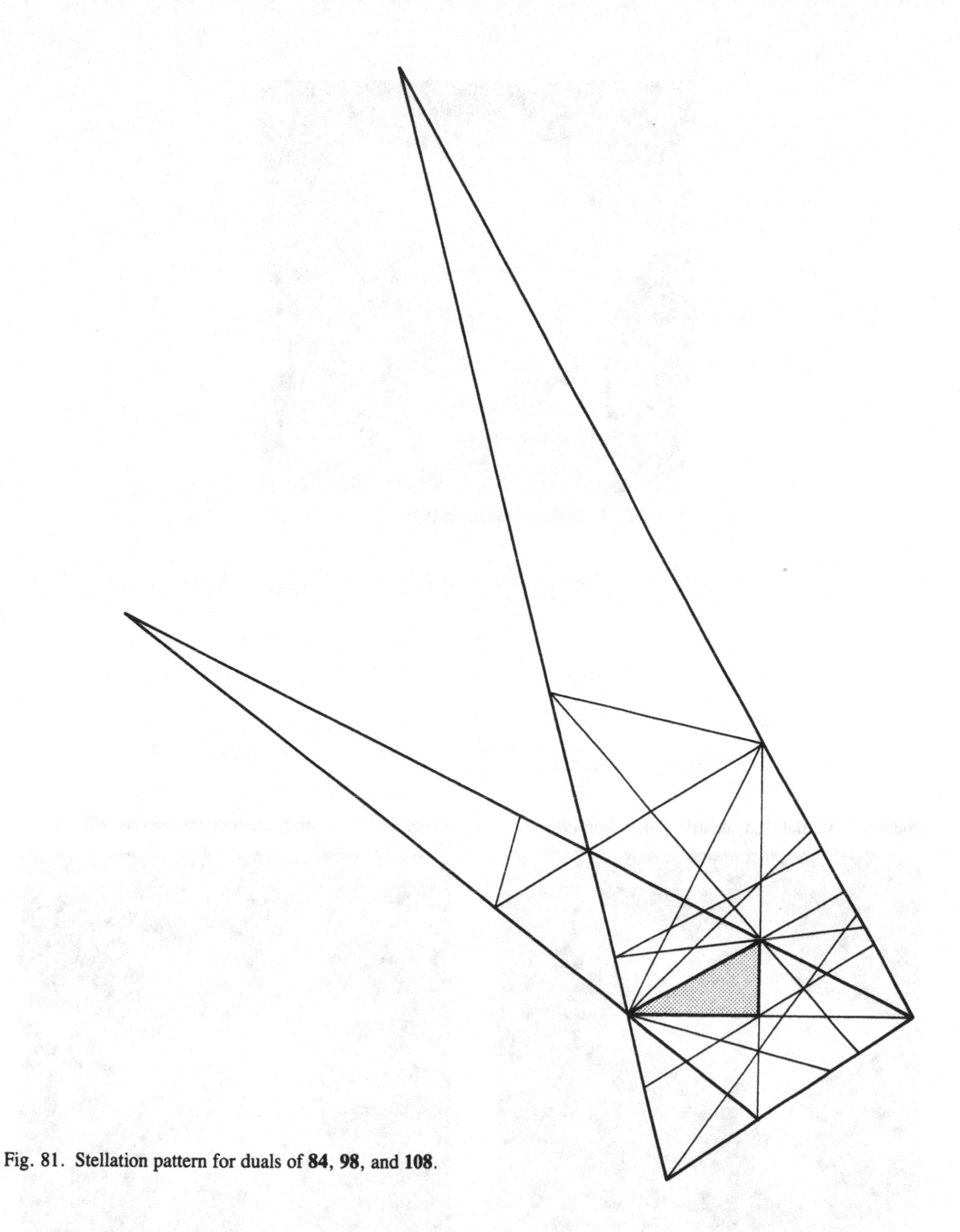

Fig. 81. Stellation pattern for duals of **84**, **98**, and **108**.

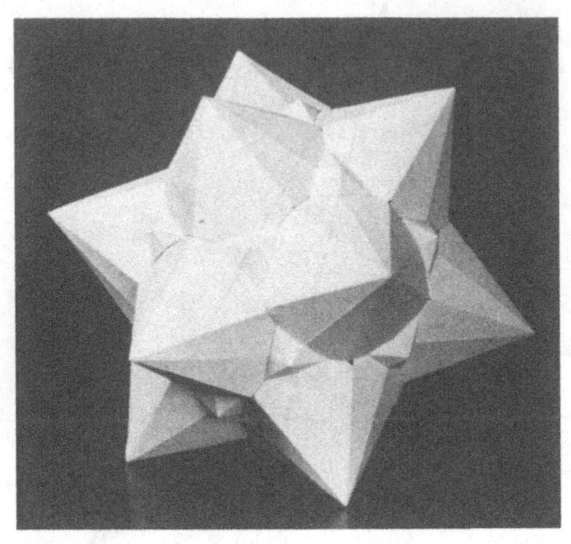

Photo 56. Tridyakisicosahedron (**84**).

Photo 57. Medial disdyakistriacontahedron (**98**).

Photo 58. Great disdyakistriacontahedron (**108**).

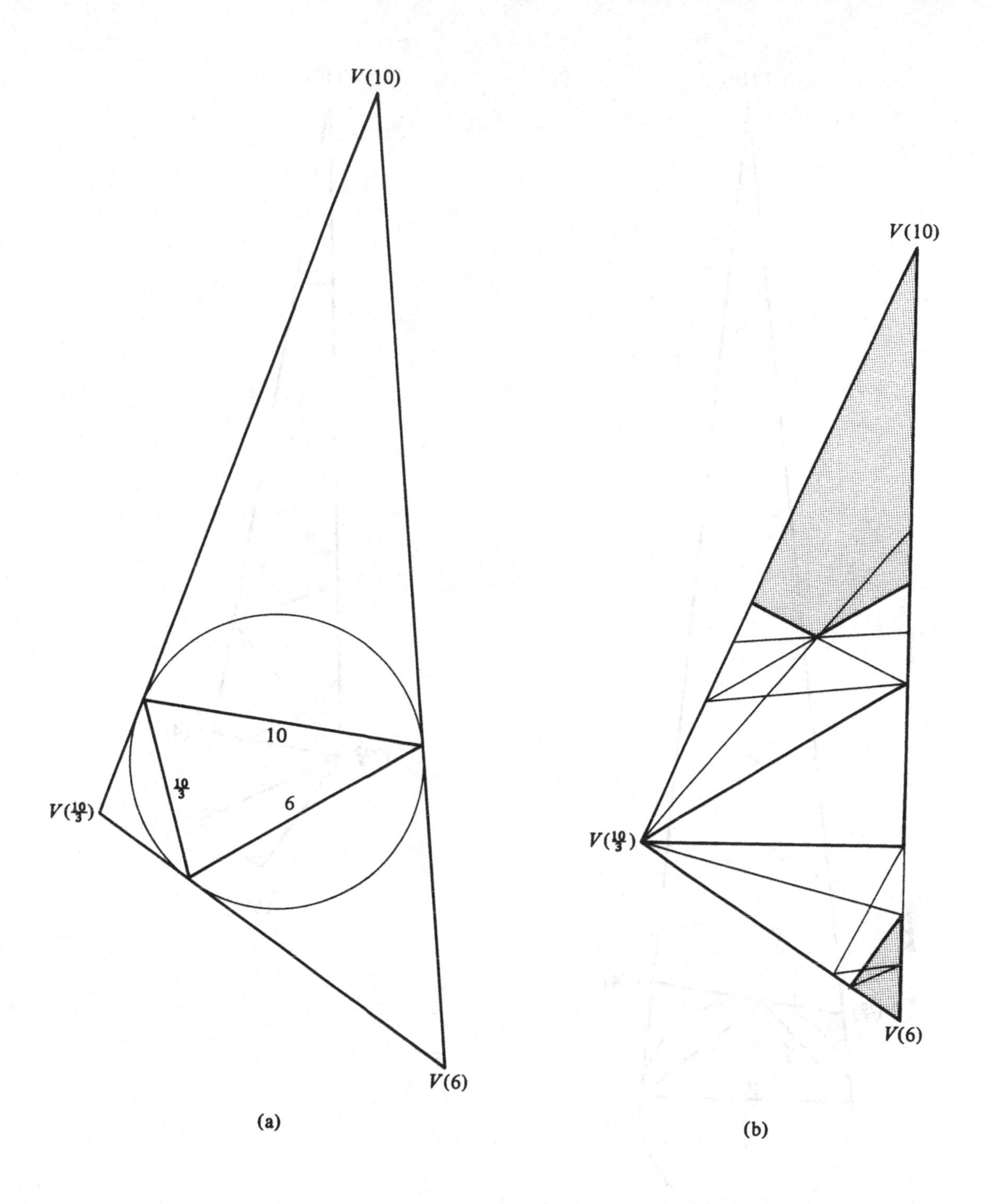

Fig. 82. Dorman Luke construction and stellation pattern for dual of **84**.

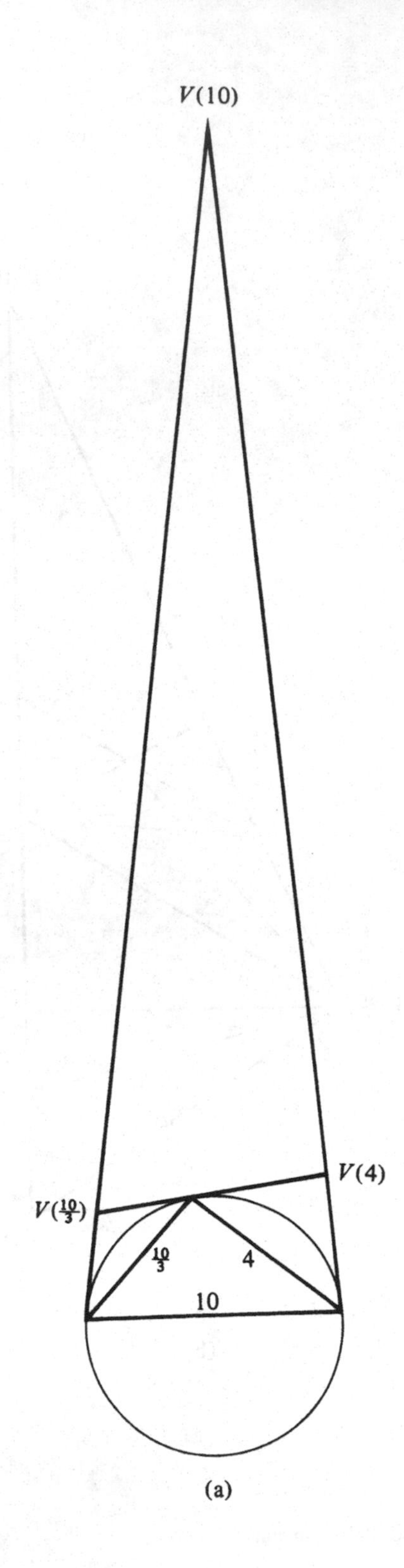

(a)

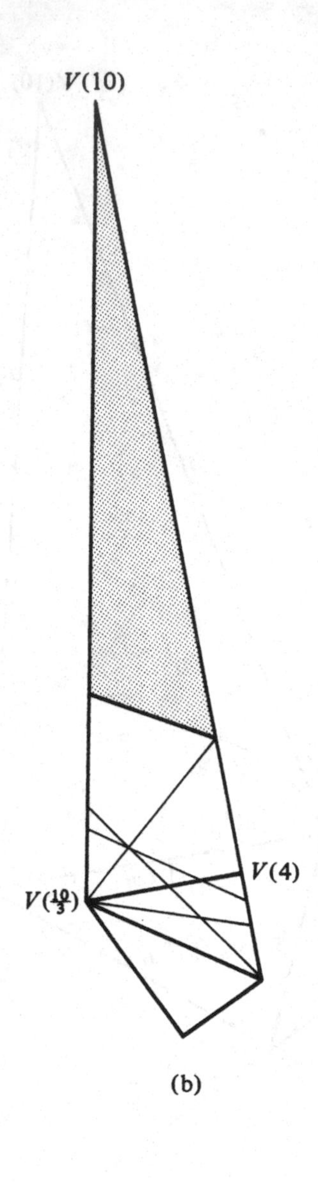

(b)

Fig. 83. Dorman Luke construction and stellation pattern for dual of **98**.

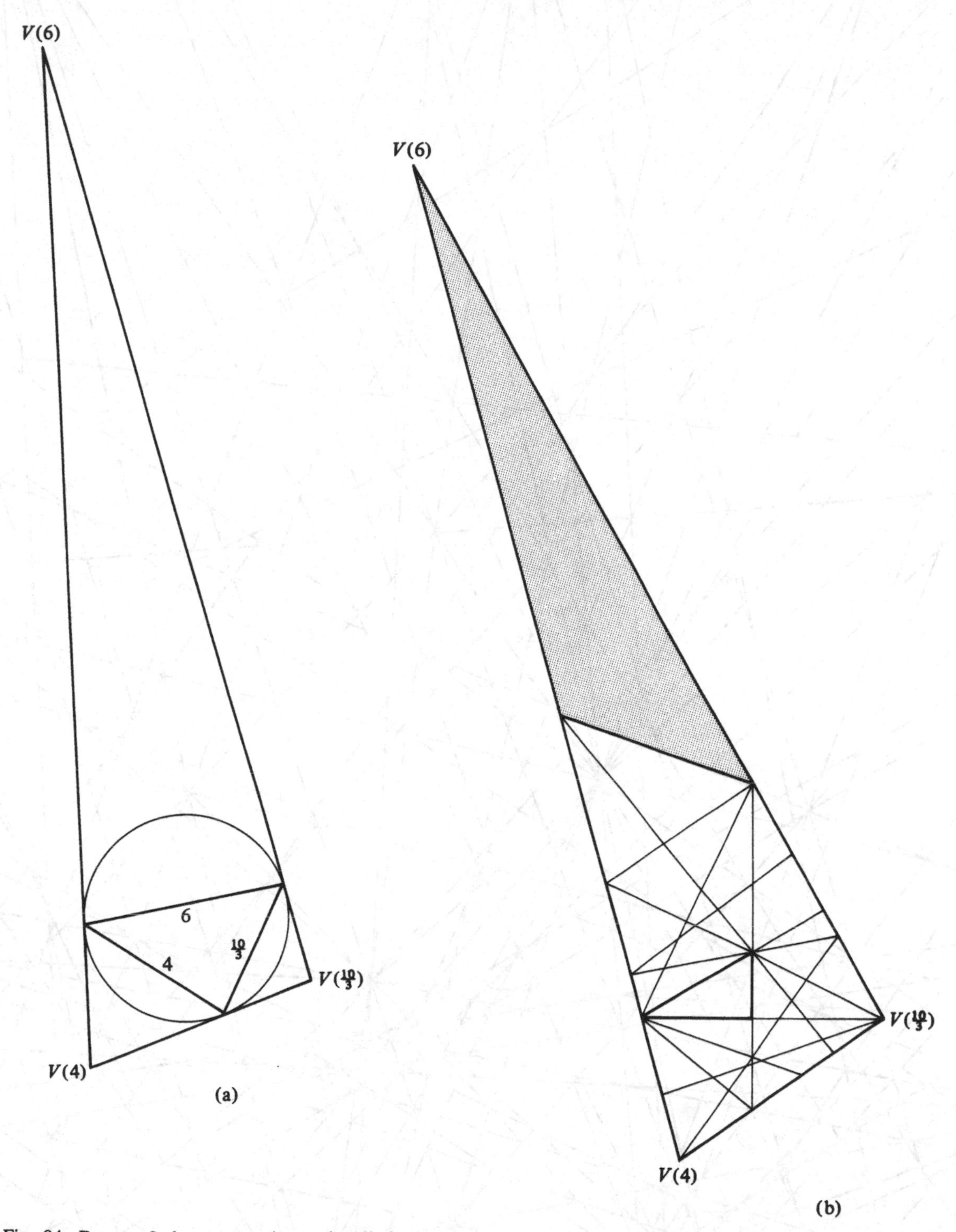

Fig. 84. Dorman Luke construction and stellation pattern for dual of **108**.

Fig. 85. Stellation pattern for a variation of the hexakisicosahedron. Angles of the central triangle: 89.218056, 59.135556, 31.646667.

These last three duals are particularly difficult to locate in a complete stellation pattern. You may judge this for yourself by just a glance at Fig. 85. This is only the inner portion of the pattern that is generated by the 120 faces of this convex core. Each face is very nearly a right-angle triangle. This poses a nearly superhuman problem to untangle the lines of this pattern. I wonder if anyone will ever find out how many basic stellation cells this has or how many stellated forms it can generate and what shape the final stellation must turn out to be. Its final vertices must lie at a great distance from the center of symmetry.

Duals of hemipolyhedra

There are nine nonconvex uniform polyhedra, the so-called hemipolyhedra, which will now be considered. They have facial planes that pass through the exact center of symmetry of the solid, and hence, mathematically speaking, they have no duals in finite space, because related vertices recede to infinity. You may have noticed in duals already presented that whenever a facial plane in the original is very near the center of symmetry, the related vertex of its dual is very far out (e.g., **97** and its dual, shown in Photo 42). The polar reciprocal formula clearly shows how the pole of a plane takes positions farther and farther from the center as the plane descends closer and closer to the center. Thus, when the plane reaches the center, the pole of this plane is at infinity.

The question may now be asked: Is there any way in which models can be made for duals of these hemipolyhedra? To answer this, look at the simplest hemipolyhedron, **67**. Its convex hull is obviously a regular octahedron. The dual of the octahedron is the cube or hexahedron. So if the dual of **67** exists, it must be a stellated form of the cube. But the cube

Photo 59. Tetrahemihexacron (**67**).

has no stellated forms, at least not in finite space. Stellation implies producing facial planes, so that their intersections generate closed cells. The facial planes of the cube can be produced, but obviously no closed cells are generated by this process. But there are what may be called open cells. In fact, there are two types of such cells, six open square prisms, one coming out from each of the six faces of the cube, and eight trihedral cells, which are shaped like open triangular pyramids without bases. These pyramids emanate one from each of the eight vertices of the cube. But the dual of **67** should really have only three tetragonal vertices and four trigonal vertices, corresponding to the three square faces and four triangular faces of **67**. This is just half the number of square prisms and triangular pyramids just mentioned. These square prisms and triangular pyramids must, of course, be imagined as having lateral faces extending to infinity, but because for model making this is impossible to construct, you need only truncate these prisms or pyramids at a convenient distance from the center of symmetry. Photo 59 shows what effect this has when it is carried out with respect to the prisms. In reality, this reduces the number of prisms to three. The model may be taken

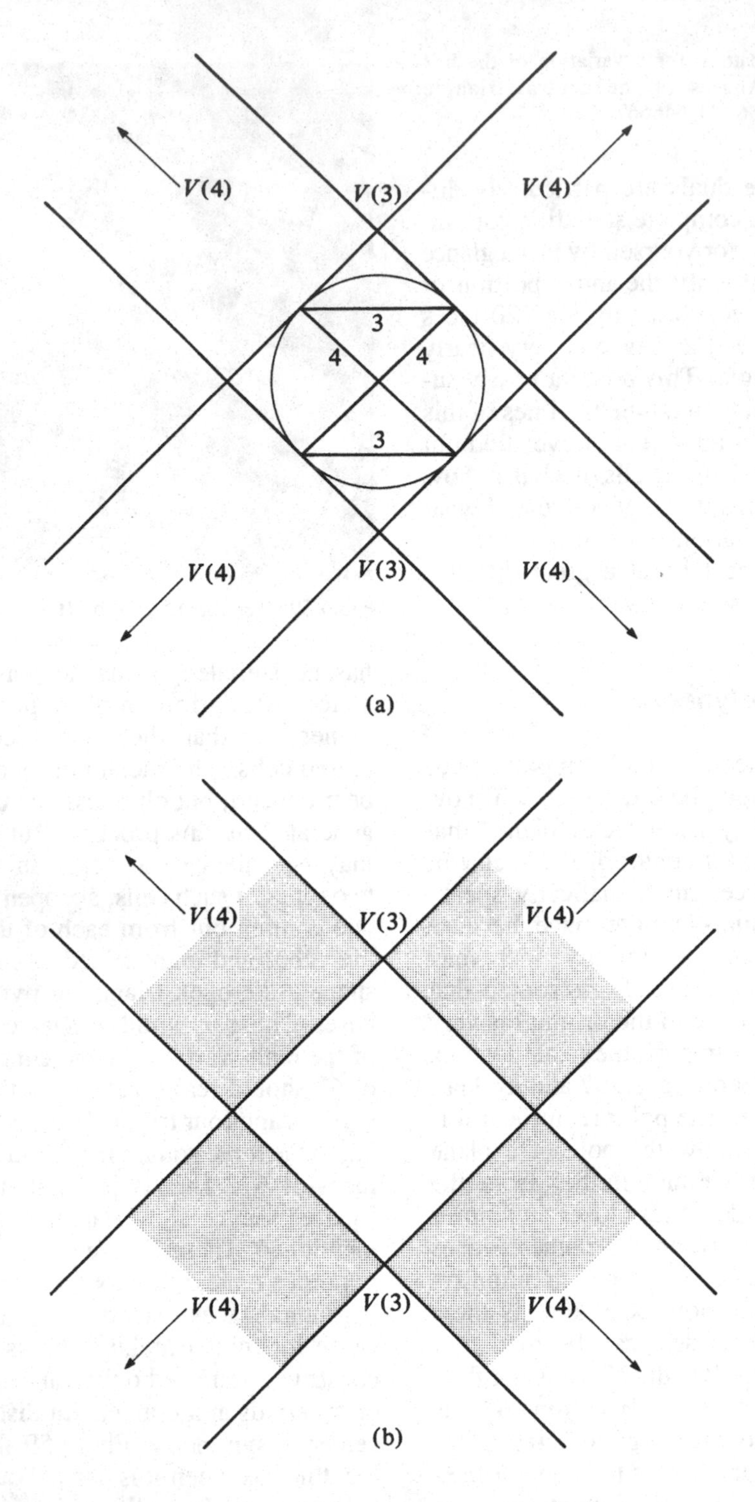

Fig. 86. Dorman Luke construction and stellation pattern for dual of **67**.

to be a compound of three square prisms interpenetrating in such a way that the inner core is simply the original cube.

How does this work out with regard to the Dorman Luke construction? Figure 86a shows this construction, and the shaded portions of Fig. 86b are the lateral faces of the square prisms. Mathematically speaking, these prisms may be thought of as infinitely tall square pyramids, the vertices of these pyramids being the required tetragonal vertices of the dual. The square faces that are the truncation planes of these pyramids are merely compromises made to satisfy the requirements of model making.

It would be possible to make a model for the dual of **67** that would use only four of the eight trihedral pyramids mentioned earlier. This would be harder to execute, but still possible as a model. These pyramids would be only vertex-connected to the inner cube, two pairs in diagonally opposite positions such as the vertices of a tetrahedron occupy when its edges are diagonals of the faces of a cube. In Fig. 86b the lateral faces of these pyramids would be the unshaded portions outside the square or, in other words, the complementary area of the shaded portions. The inner square is a face of the cube serving as the core of the stellation process.

A close examination of the polygon that is generated from the Dorman Luke construction reveals that mathematically the situation is more correctly shown as in Fig. 87. This shows a quadrilateral with two vertices in finite space, the $V(3)$ vertices, and the other two, the $V(4)$ vertices, at infinity. You can trace out the four sides of this strange quadrilateral, still a crossed polygon in one sense, by starting from a $V(3)$ vertex and going to a $V(4)$ vertex at infinity in one direction (shown by the arrow) and returning from infinity from the opposite direction to arrive at the opposite $V(3)$ vertex. A repetition of this process brings you from that

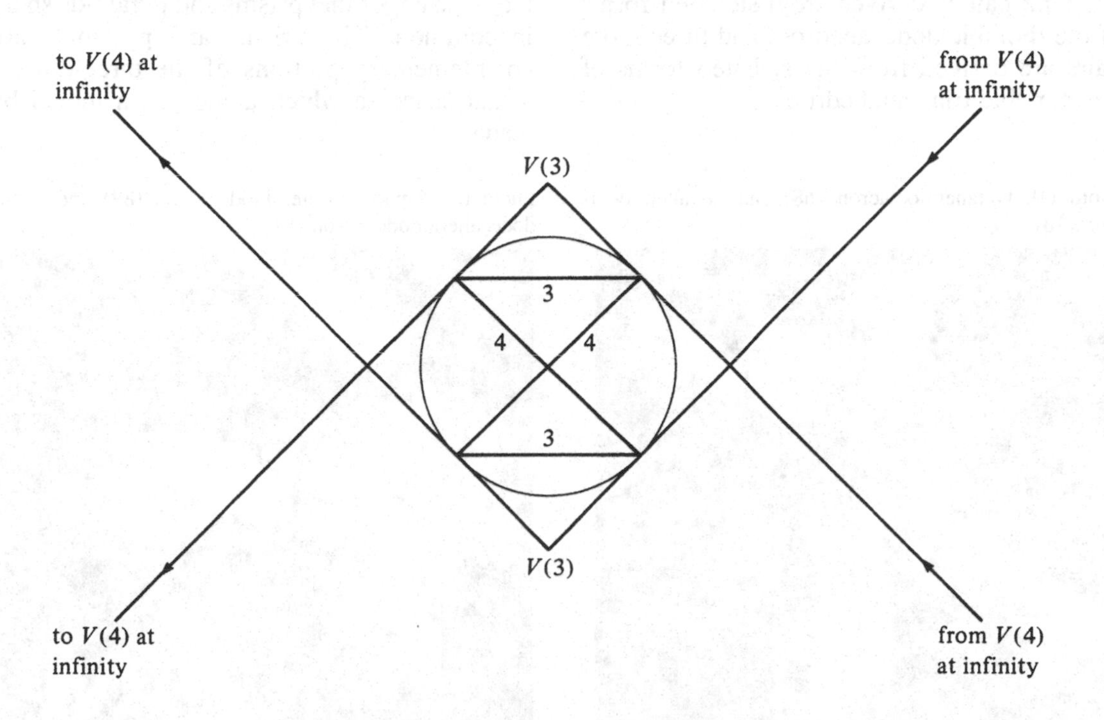

Fig. 87. Facial plane for dual of **67**, with two vertices at infinity.

Table 5. *Duals of hemipolyhedra*

Polyehdron	Convex hull	Dual of hull	Dual
67	Octahedron	Cube	Photo 59
68, **78**	Cuboctahedron	Rhombic dodecahedron	Photo 60
89, **91**	Icosidodecahedron	Rhombic triacontahedron	Photo 61
100, **102**			Photo 62
106, **107**			Photo 63

point back to the starting point. The lines going out to infinity and returning from infinity must be thought of as being two lines and counted as two sides of the quadrilateral.

The prism method has been chosen for the model shown in Photo 59 because it is easier to execute and because it can be extended to the rest of the hemipolyhedra with greater success. These are set out in Table 5. The placement of hidden vertices causes these duals to arrange themselves in distinct pairs, except for **67**. One pair is derived from stellated forms of the rhombic dodecahedron, and three more pairs are derived from the stellated forms of the rhombic triacontahedron.

Figures 88 through 95 show the Dorman Luke construction and the shaded portions of respective stellation patterns from which models of the duals can be made. These are shown in Photos 60 through 63. Figure 96 shows the interesting interrelationship of the duals derived from stellations of the rhombic triacontahedron. In all of these, the facial planes are really quadrilaterals of the same type as shown in Fig. 87. These facial planes, therefore, do not appear on the models, but only the same edges that prisms and pyramids share in common. The prisms and pyramids are complementary portions of the three-dimensional space in which alone can a model be made.

Photo 60. Octahemioctacron (**68**) and hexahemioctacron (**78**).

Photo 61. Small icosihemidodecacron (**89**) and small dodecahemidodecacron (**91**).

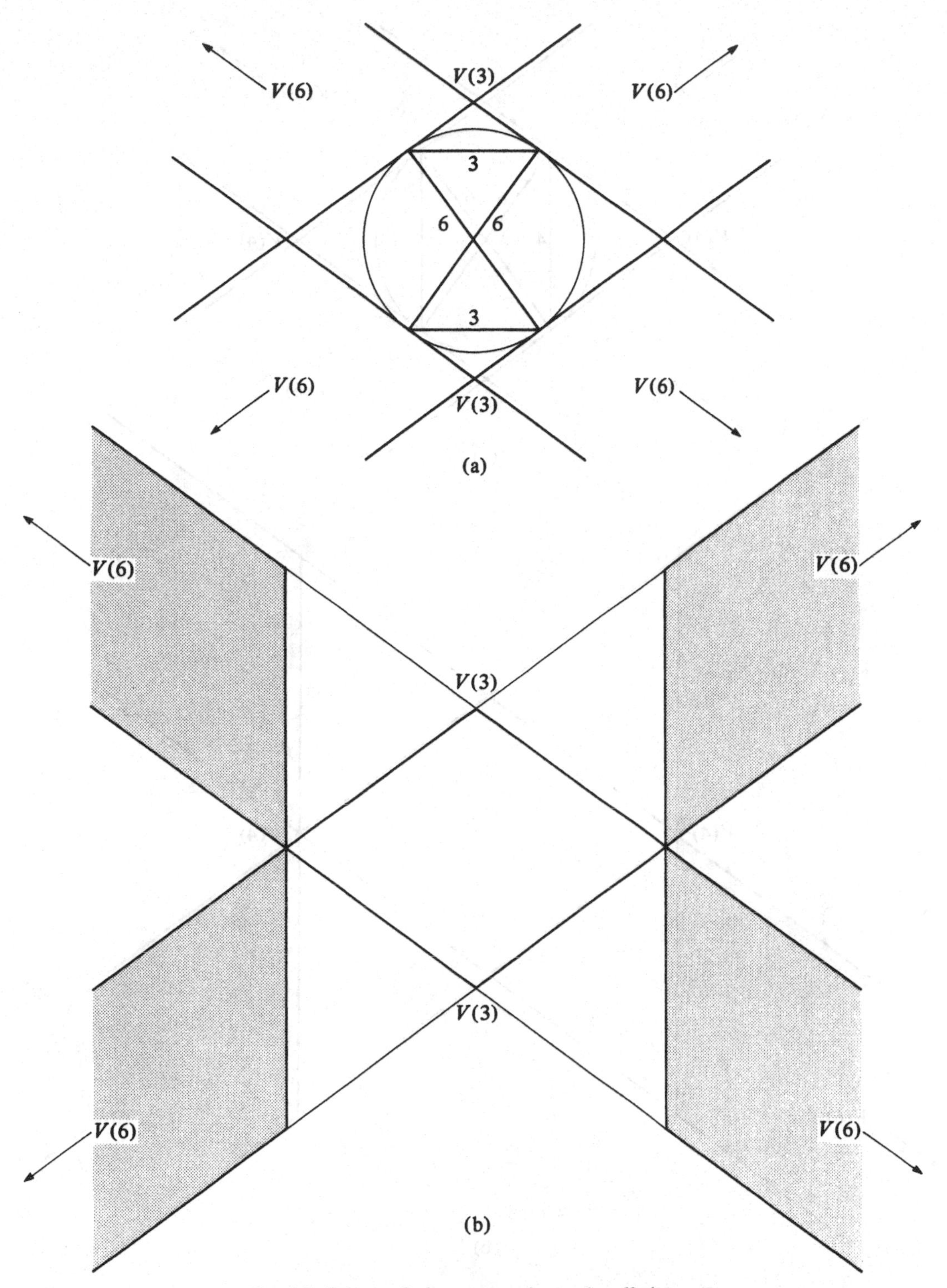

Fig. 88. Dorman Luke construction and stellation pattern for dual of **68**.

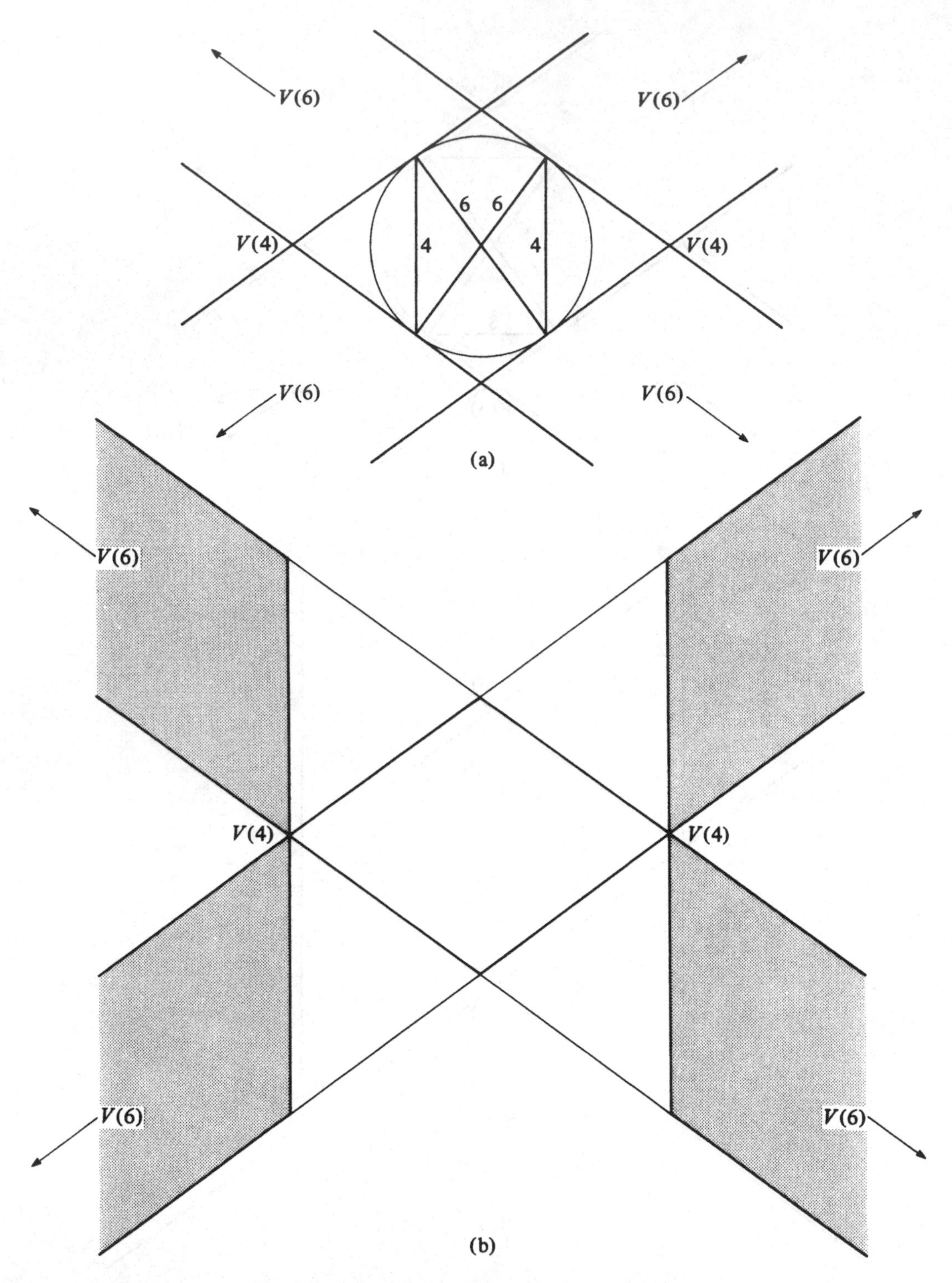

Fig. 89. Dorman Luke construction and stellation pattern for dual of **78**.

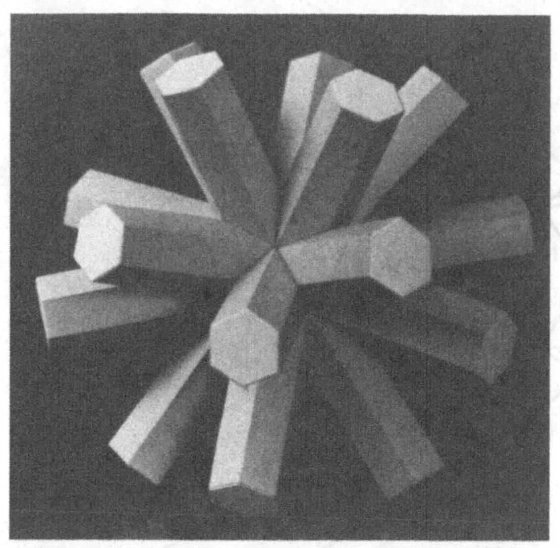

Photo 62. Small dodecahemicosacron (**100**) and great dodecahemicosacron (**102**).

My investigation of these duals for the hemipolyhedra, models that I admit are not particularly interesting or attractive in themselves, led me to discover what I would call completely new polyhedral forms. These new forms arise as an extrapolation that goes beyond the stage called the final stellation of a convex polyhedron. The final stellation is always unique, because it utilizes all the finite cells generated by the intersections of all the facial planes. But if some of the cells are allowed to be open and then are truncated at a convenient distance from the center of symmetry, as was done with the hemipolyhedra, a form emerges that I would call a stellation to infinity.

Photo 63. Great icosihemidodecacron (**106**) and great dodecahemidodecacron (**107**).

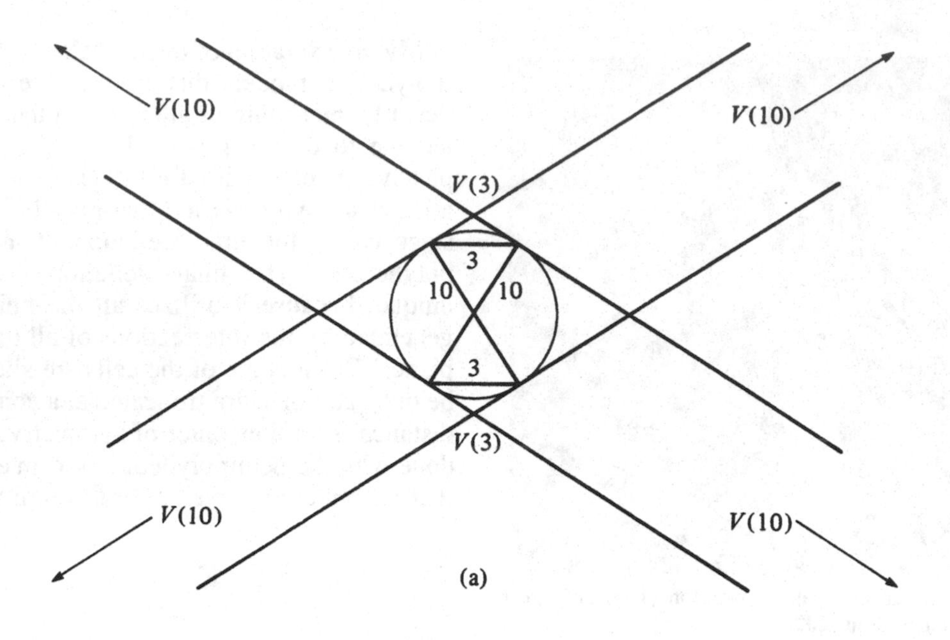

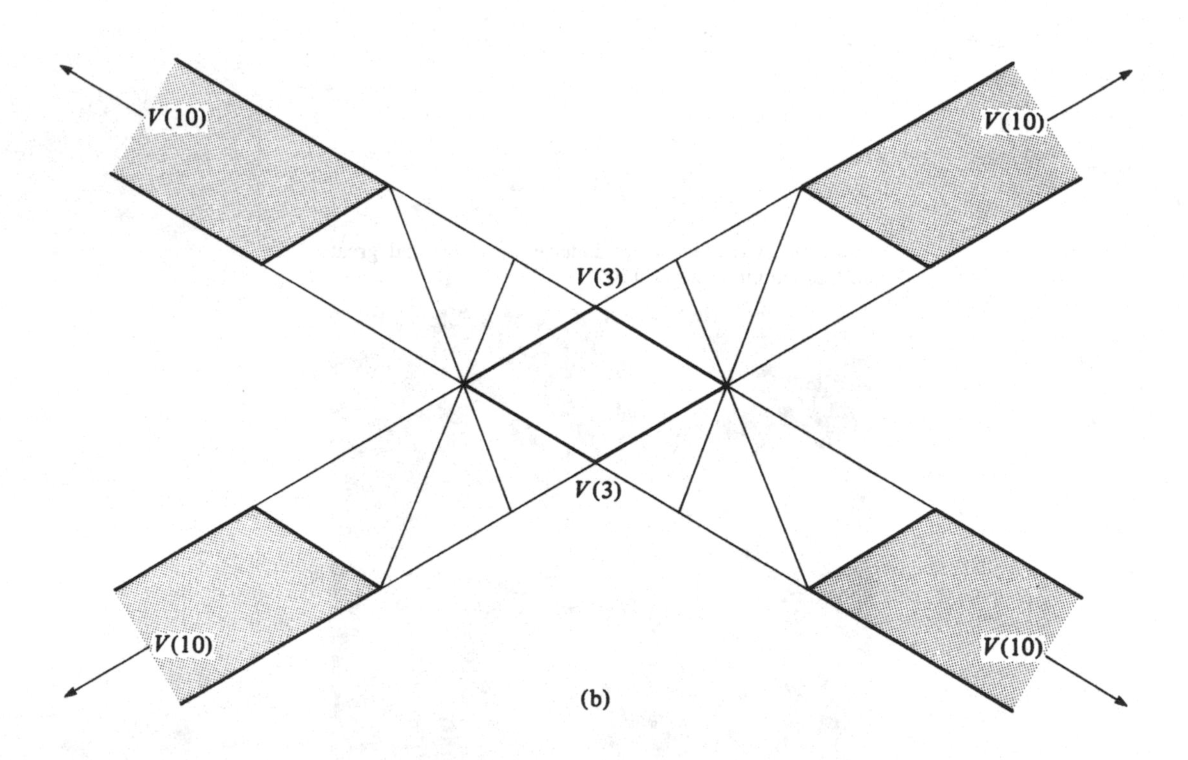

Fig. 90. Dorman Luke construction and stellation pattern for dual of **89**.

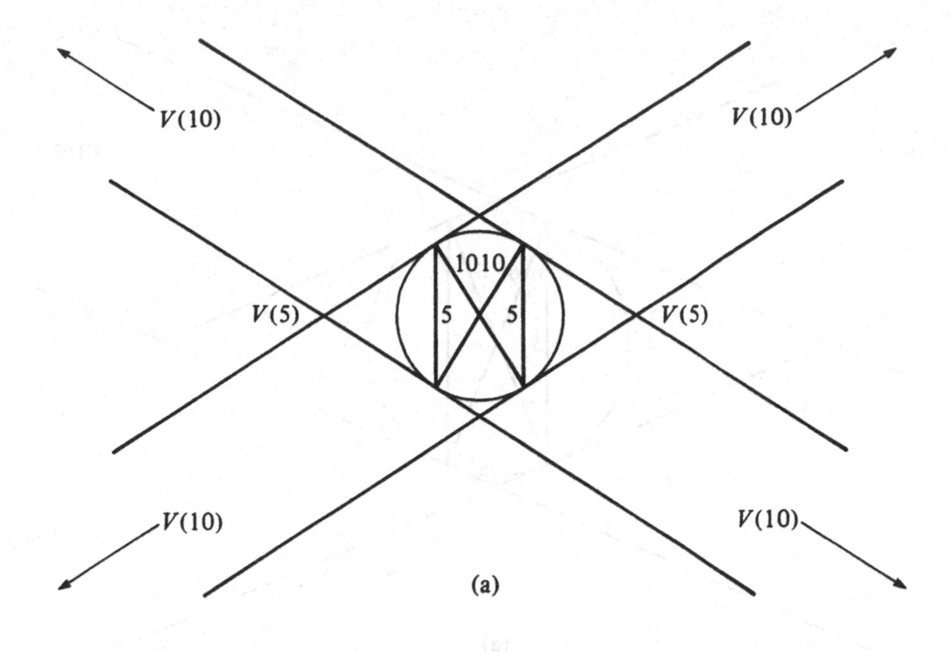

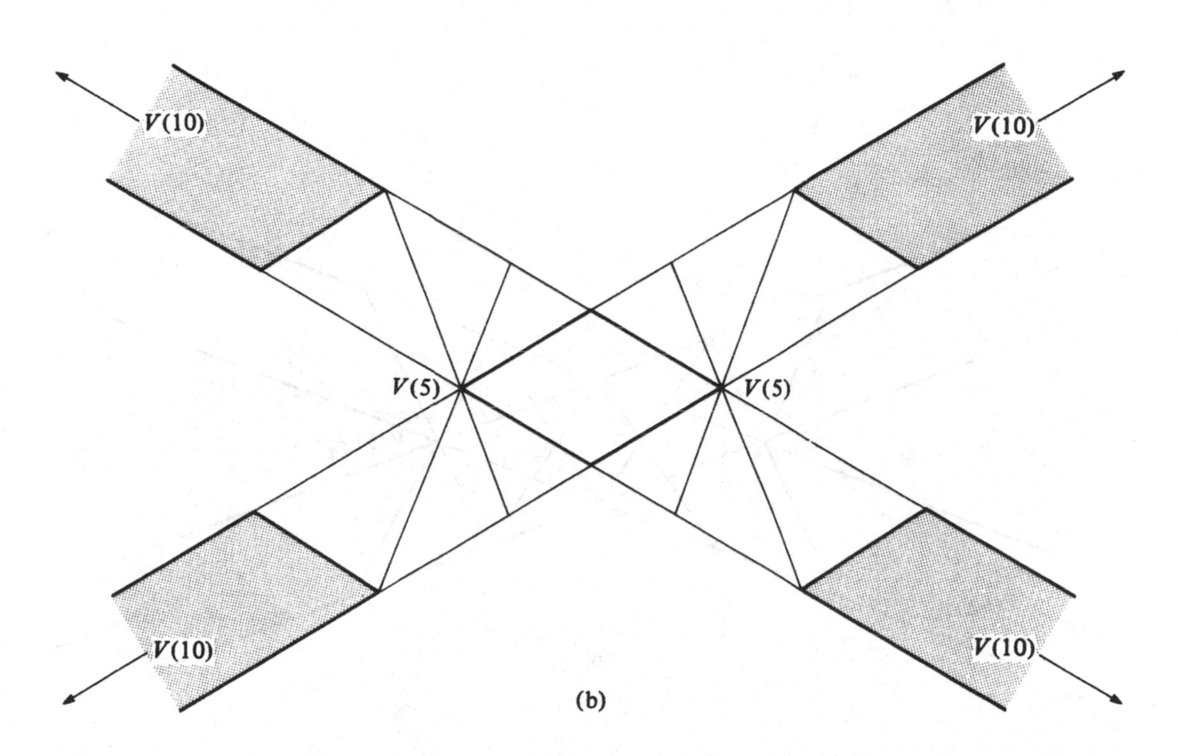

Fig. 91. Dorman Luke construction and stellation pattern for dual of **91**.

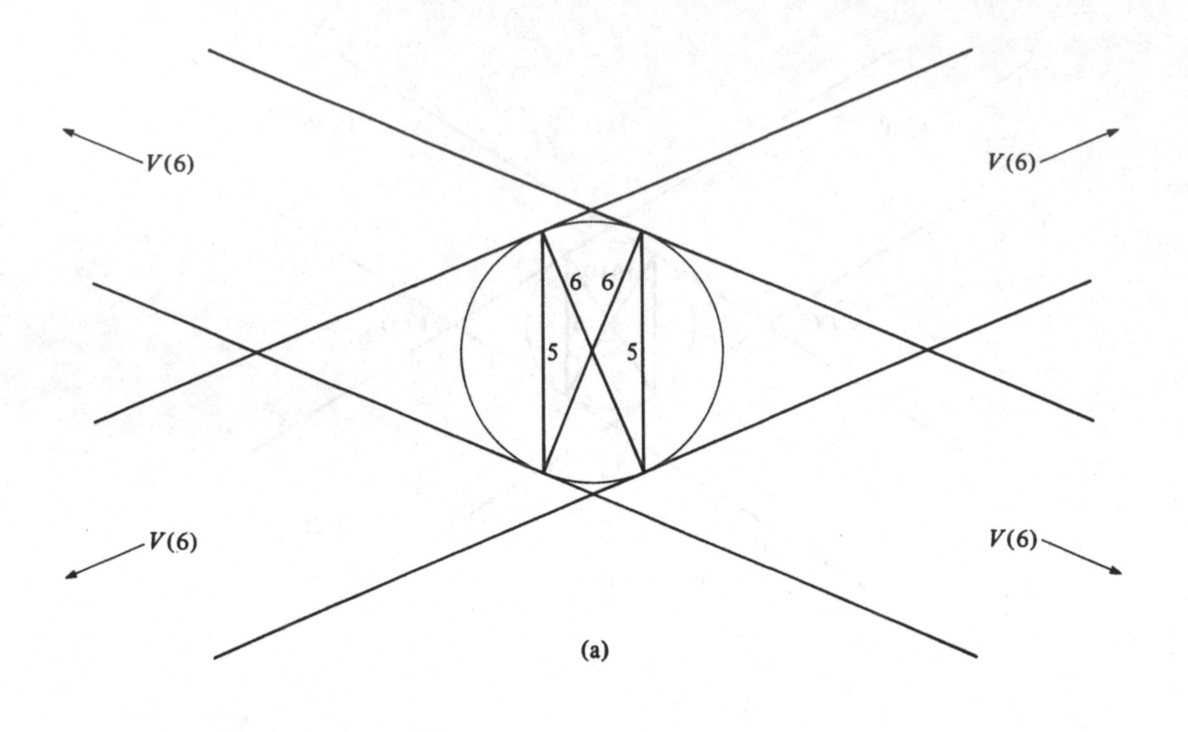

(a)

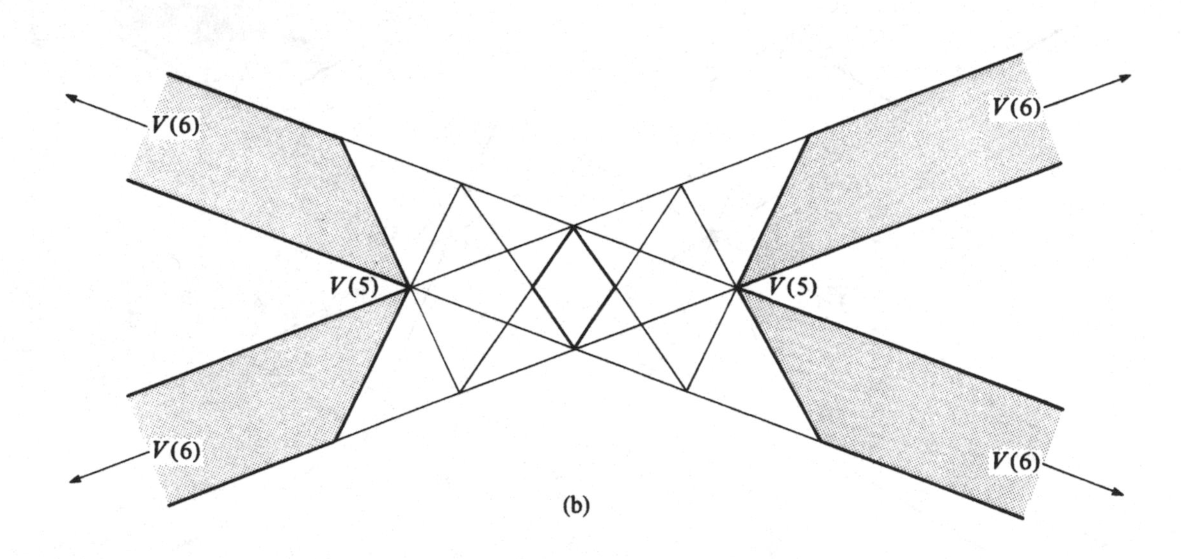

(b)

Fig. 92. Dorman Luke construction and stellation pattern for dual of **100**.

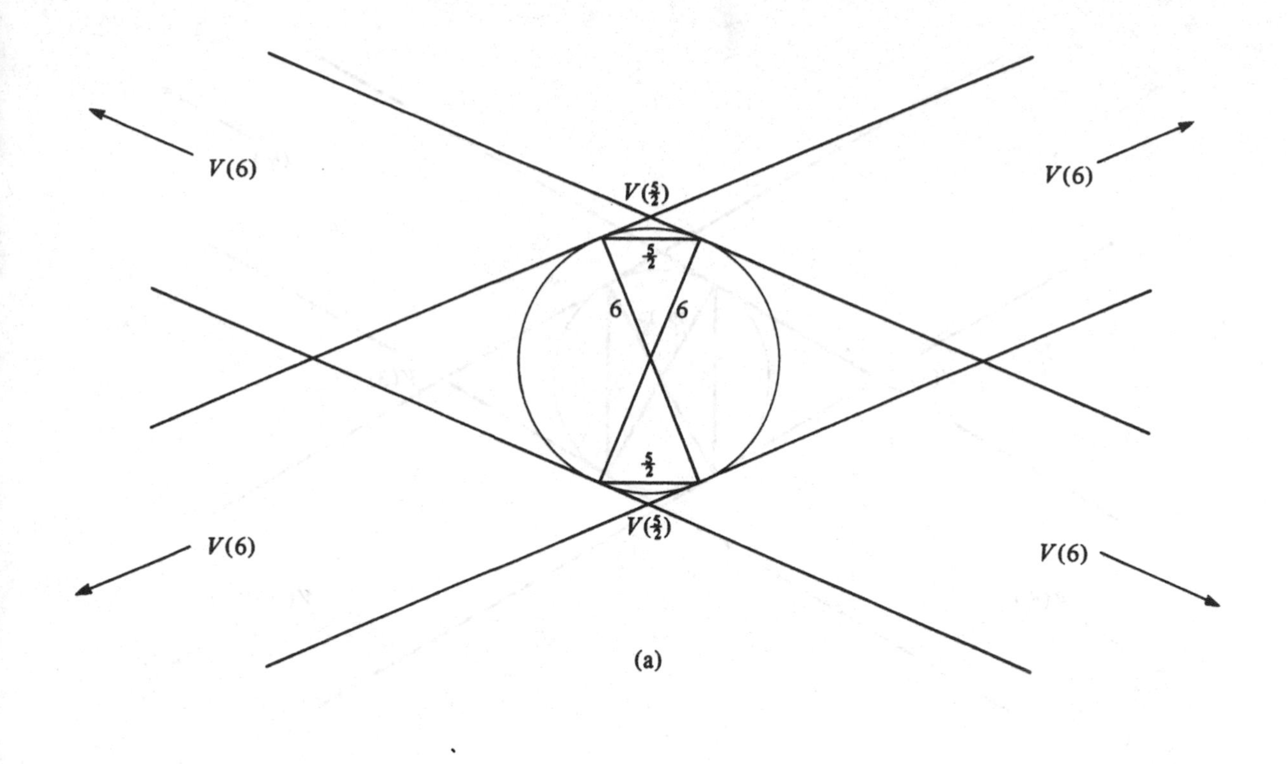

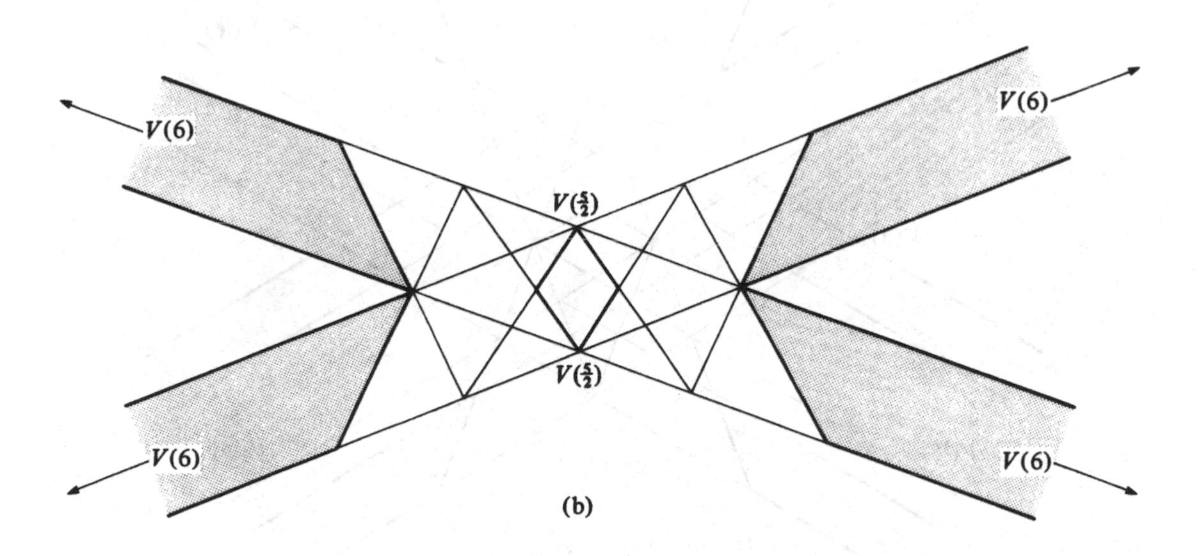

Fig. 93. Dorman Luke construction and stellation pattern for dual of **102**.

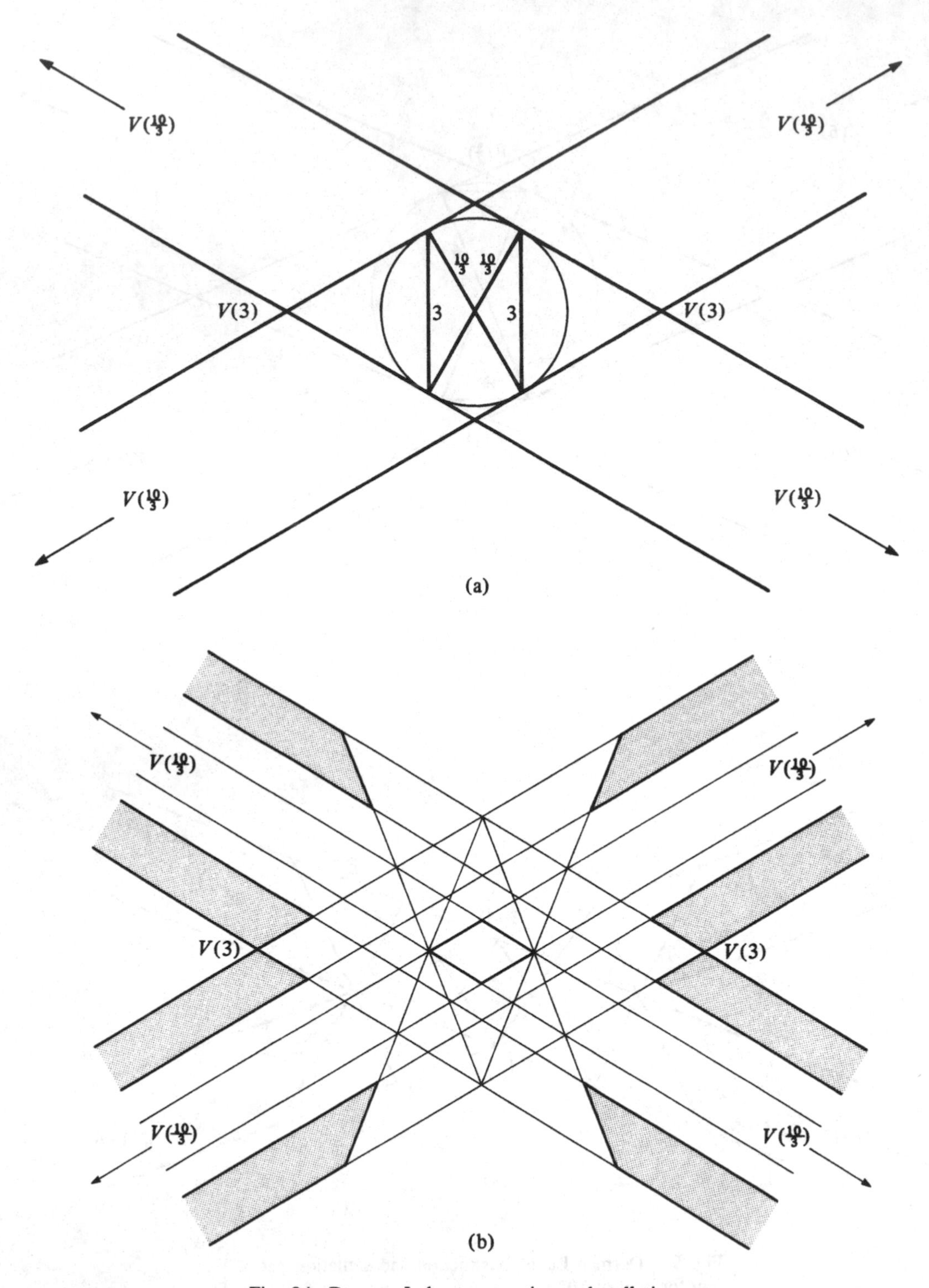

Fig. 94. Dorman Luke construction and stellation pattern for dual of **106**.

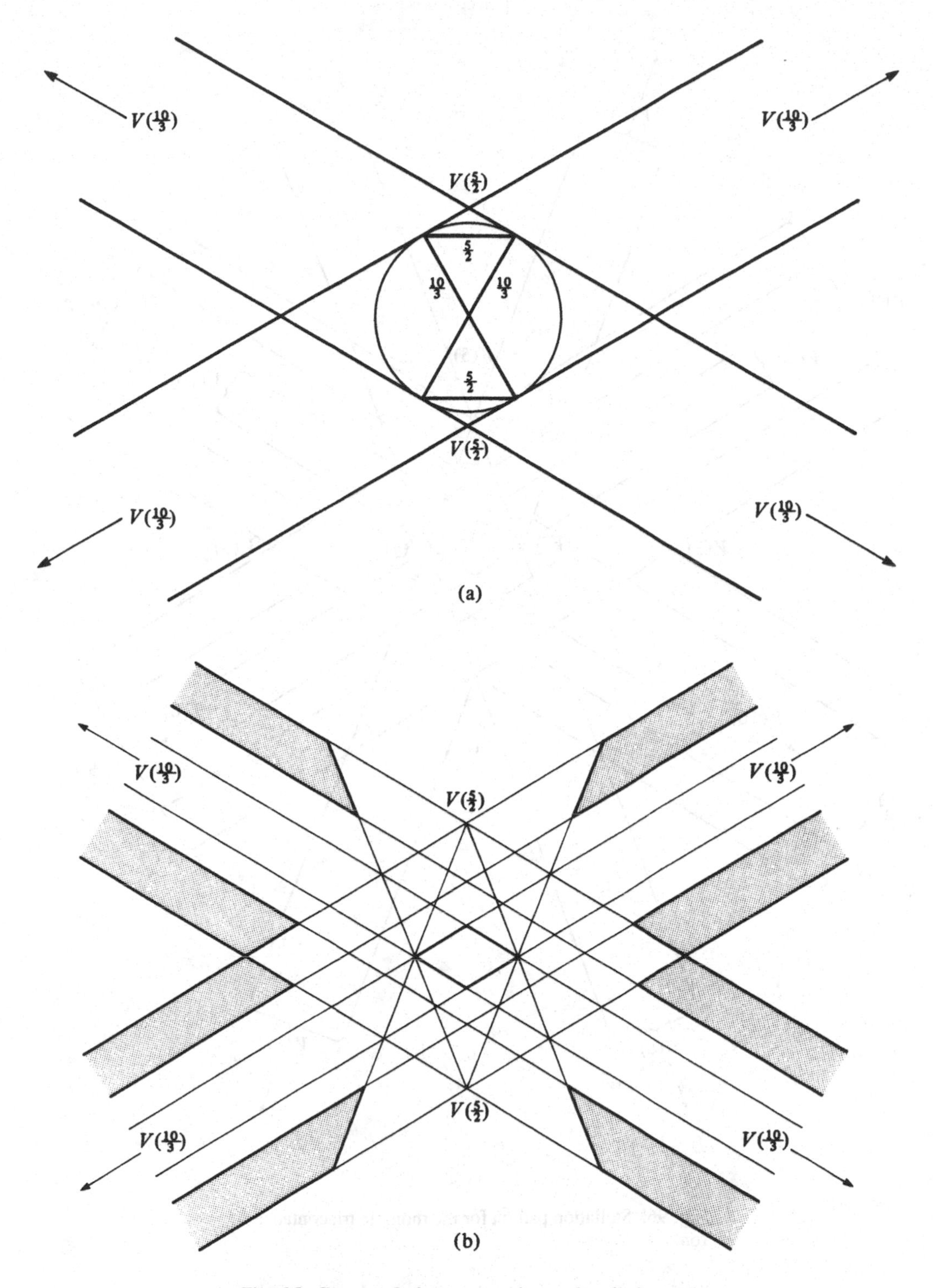

Fig. 95. Dorman Luke construction and stellation pattern for dual of **107**.

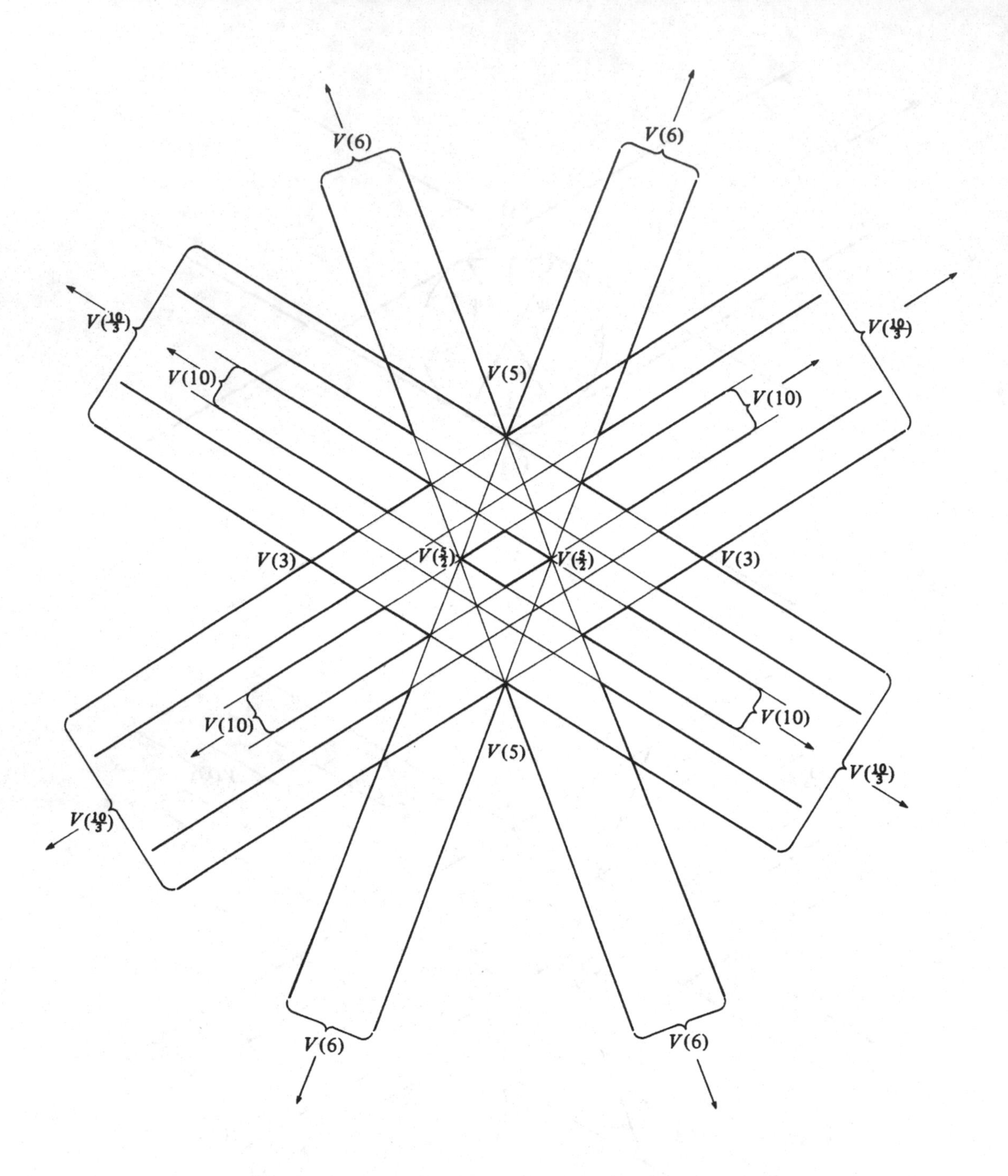

Fig. 96. Stellation pattern for the rhombic triacontahedron.

Photo 64. Stellation to infinity, rhombic dodecahedron.

Photo 65. Stellation to infinity, five cubes.

Photo 64 shows what happens when this method is applied to the rhombic dodecahedron. Figure 97 shows the stellation pattern for this. When the same method is applied to the rhombic triacontahedron, some fantastically beautiful shapes emerge. Photo 65 shows the stellation to infinity for the compound of five cubes. This is a compound of five interpenetrating sets of prisms of the kind shown as the dual of **67** in Photo 59. Photo 66 is truly remarkable. Here the polyhedral equivalent of a fireworks display manifests itself in the final stellation to infinity of the rhombic triacontahedron. In color these models become even more attractive. The complete stellation pattern to infinity is shown in Fig. 98. In Photo 66, the squares, the hexagrammic or double-triangle planes, and the decagrammic or double-star planes all seem to be caught by the camera at just one stage of their journey outward toward infinity. The model seems to suggest an explosion in progress rather than a fixed form.

This new world of polyhedral shapes is one of those beckoning side paths that could easily become a major highway for further investigation. It is one that you are invited to explore on your own.

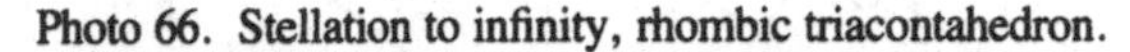

Photo 66. Stellation to infinity, rhombic triacontahedron.

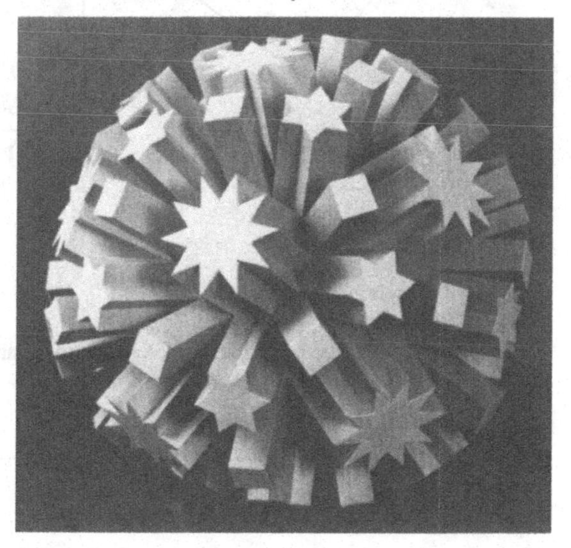

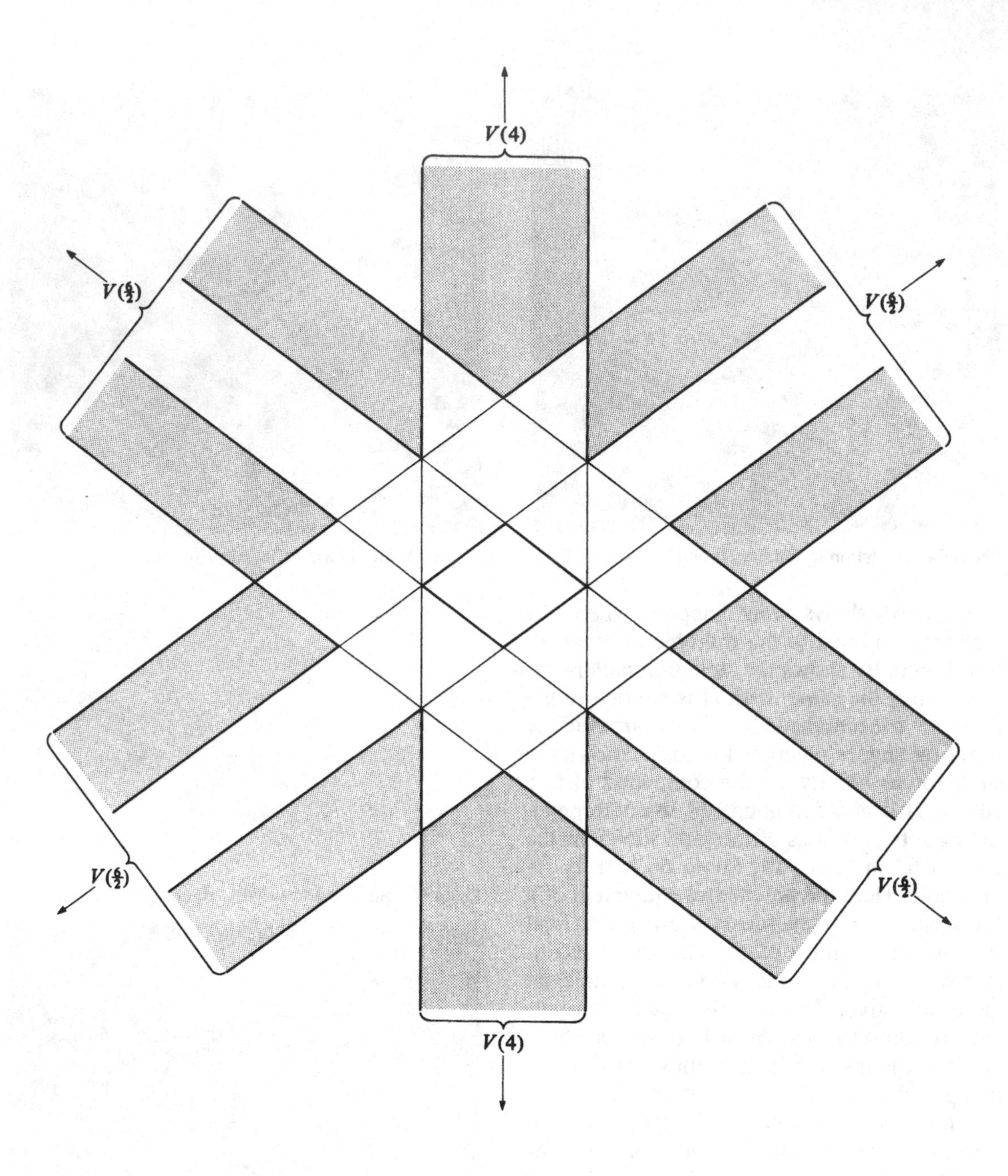

Fig. 97. Stellation to infinity: the rhombic dodecahedron.

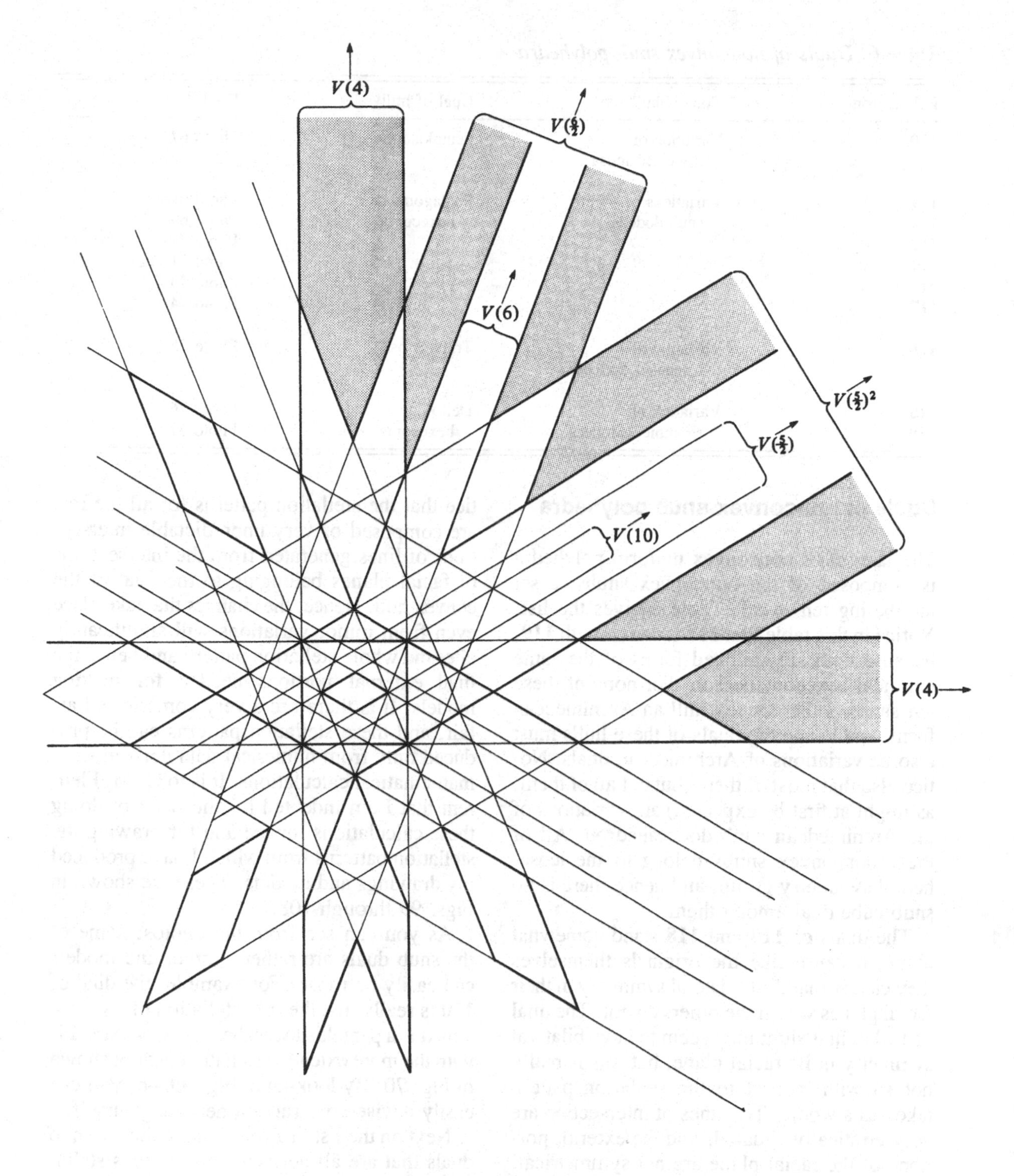

Fig. 98. Stellation to infinity; the rhombic triacontahedron.

Table 6. *Duals of nonconvex snub polyhedra*

Polyhedron	Convex hull	Dual of hull	Dual
110	Variation of truncated icosa	Pentakisdodeca	Photo 67
111	Variations of snub dodeca	Pentagonal hexeconta	Photo 68
112			Photo 69
113			Photo 70
114			Photo 71, 72
116			Photo 73
117			Photo 74
118	Variation of truncated dodeca	Triakisicosa	Photo 75
115	Variation of rhombicosidodeca	Deltoidal hexeconta	Photo 76
119			Photo 77

Duals of nonconvex snub polyhedra

The last set of nonconvex uniform polyhedra is composed of the nonconvex snubs, a set numbering ten in all. Table 6 gives the list. Notice in this table that **115** is paired with **119**, because both are stellated forms of the same deltoidal hexecontahedron. For none of these ten snubs is the convex hull an Archimedean form, and hence the duals of these hulls must also be variations of Archimedean duals. Notice also that most of them (but not all of them, as might at first be expected) are variations of the Archimedean snub dodecahedron. All of these nonconvex snubs belong to the icosahedral symmetry group, and hence there is no snub cube dual among them.

The duals of **110** and **118** stand somewhat alone, because like the originals themselves they clearly manifest bilateral symmetry in their facial planes which the others do not. The dual of **113** at first sight may seem to have bilateral symmetry in its facial plane, but this is really not so with respect to the stellation pattern taken as a whole. The lines of intersection are very erratically situated, and the exterior portions of the facial plane are not symmetrical, whereas for **110** and **118** even the exterior portions show bilateral symmetry. Finally, notice that the stellation patterns for all the rest are composed of very unpredictable intersections of lines generated from the intersections of facial planes belonging to the dual of the convex hull. Hence, the changes that take place even from minor variations will significantly alter the whole stellation pattern and hence the final external portions needed for making models. It is therefore a very complicated affair, and these stellation patterns can be produced only from numerical data derived from mathematical calculations. It is to G. M. Fleurent that I am indebted for the work of doing these calculations for me and for drawing the stellation patterns from which I have produced my drawings and models. These are shown in Figs. 99 through 108.

As you can see from the photos, some of the snub duals are rather simple, and models can easily be made. For example, the dual of **110** is really just the first stellation of its core, which is a pentakisdodecahedron. Compare this with the more extended stellation pattern shown in Fig. 70. By looking at Fig. 99a–b, you can easily devise construction nets for yourself.

Next on the list in Table 6 come the six snub duals that are all derived from various stellations of a pentagonal hexecontahedron. Photo 68 shows a model actually derived from the

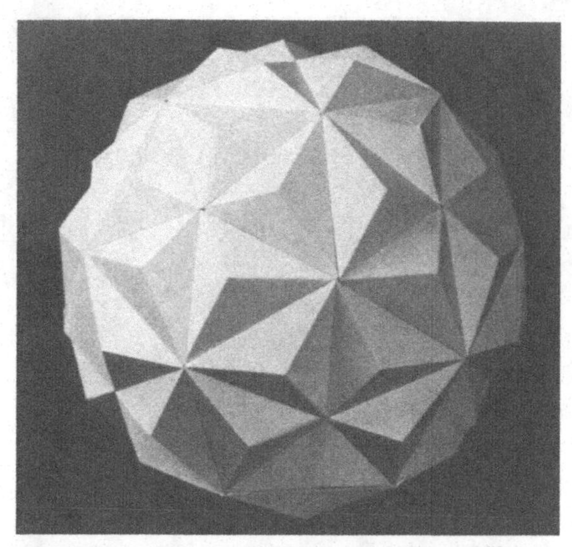

Photo 67. Small hexagonal hexecontahedron (**110**).

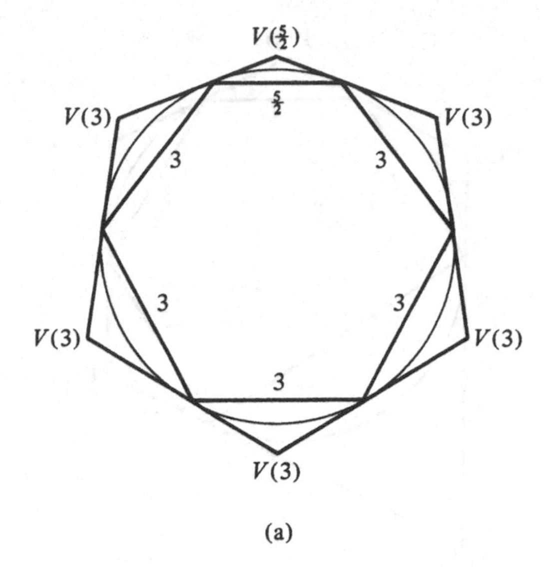

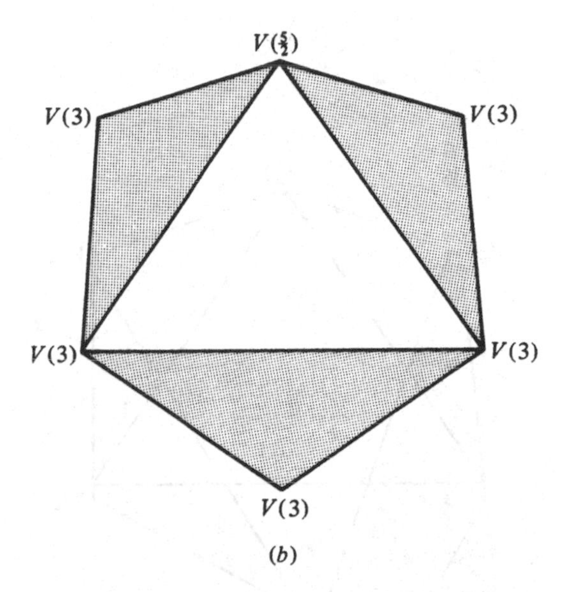

Fig. 99. Dorman Luke construction and stellation pattern for dual of **110**.

Archimedean form. The stellation pattern is shown in Fig. 100c. If you compare this with Fig. 100b, which is the mathematically correct pattern, you see a very tiny shaded triangular portion at one $V(3)$ vertex that should be connected with the two shaded triangular portions at the two other $V(3)$ vertices and with the little hooked portion associated with the base of the shaded portion of the $V(5)$ vertex. On a large scale, such a model could quite easily be made. It will be left for you to attempt this, if you so wish.

The next model is the dual of **112**, shown in Photo 69. This also is still a very simple model to make. Figure 101a–b shows the required drawings. Notice that the vertex figure of **112** is a crossed polygon of six sides, and the face of its dual is a nonconvex polygon of six sides, but with four of its vertices hidden inside the model. Notice, too, how the core of this stellation has rather noticeably shifted away from bilateral symmetry, and with it the whole stellation pattern.

For a model of the dual of **113**, shown in Photo 70, some slight compromises are again introduced. Figure 102a shows that both the vertex figure of **113** and hence also the facial plane of its dual are convex pentagons, nearly regular, but not quite so. The vertex figure has four equal sides, and hence the facial plane has four equal angles, but only three equal sides of one length and the other two of a slightly shorter length. The compromises mentioned are made visibly explicit in Fig. 102b, showing the stellation pattern, with one enlargement area indicating where this compro-

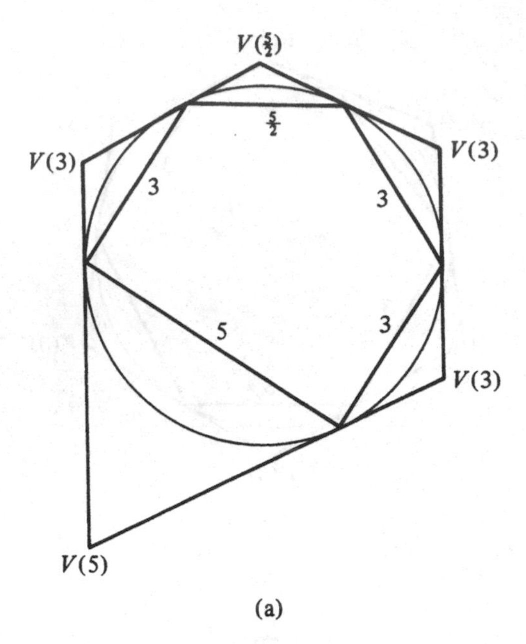

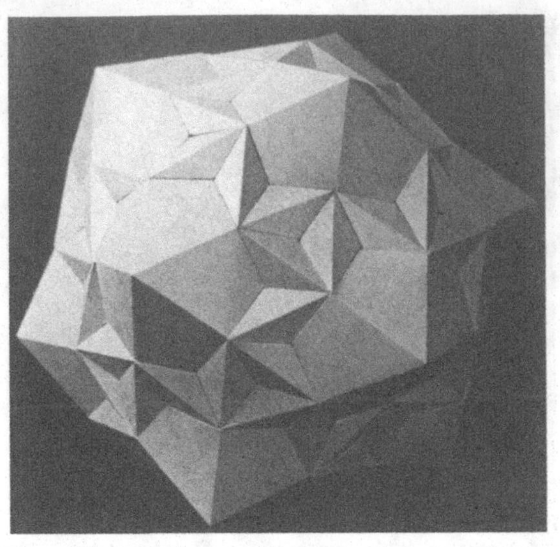

Photo 68. Medial pentagonal hexecontahedron (**111**).

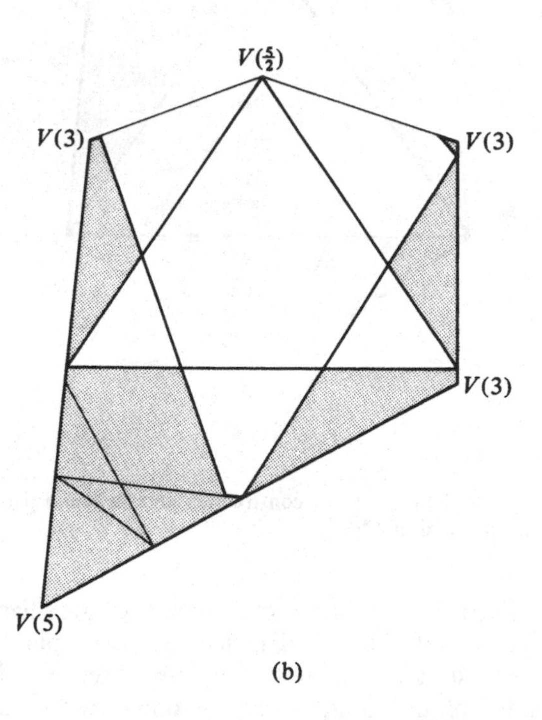

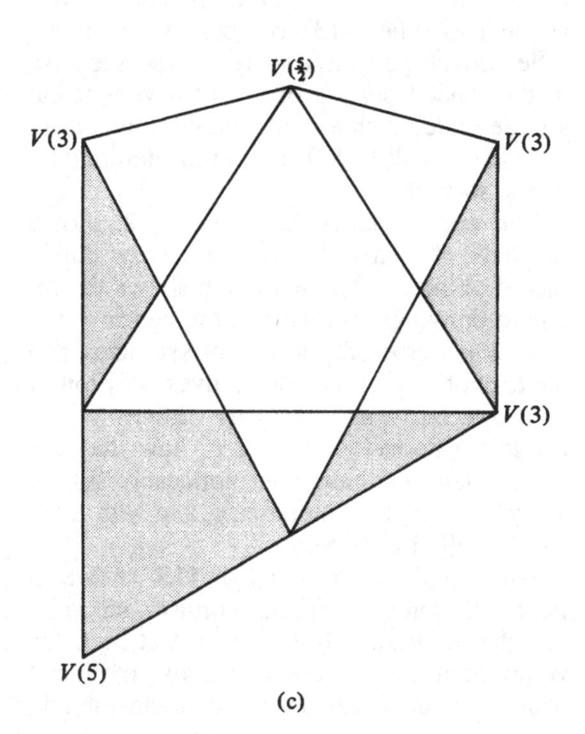

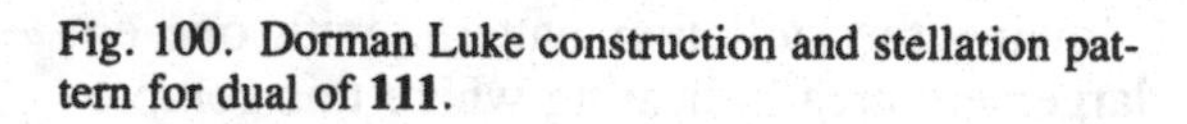

Fig. 100. Dorman Luke construction and stellation pattern for dual of **111**.

Photo 69. Medial hexagonal hexecontahedron (**112**).

mise is made. Interestingly enough, this pattern also shows that none of the vertices of this dual are hidden vertices, but some rather small triangular shapes appear along the sides of this pattern that belong to wedge-shaped cells along the edges of the dual model. Because of the complicated way in which all of the shaded areas of Fig. 102b are interconnected, you may look at Fig. 102c for patterns or nets from which a model can be made. Each of these three nets is, in effect, a separate vertex part of the dual model. Hence, you will need 60 + 20 + 12 of these parts. The assembly is also very complicated, and so all I can do by way of commentary is to invite you to attempt it. It calls for a great deal of patience to complete it, but it does make a very rigid model.

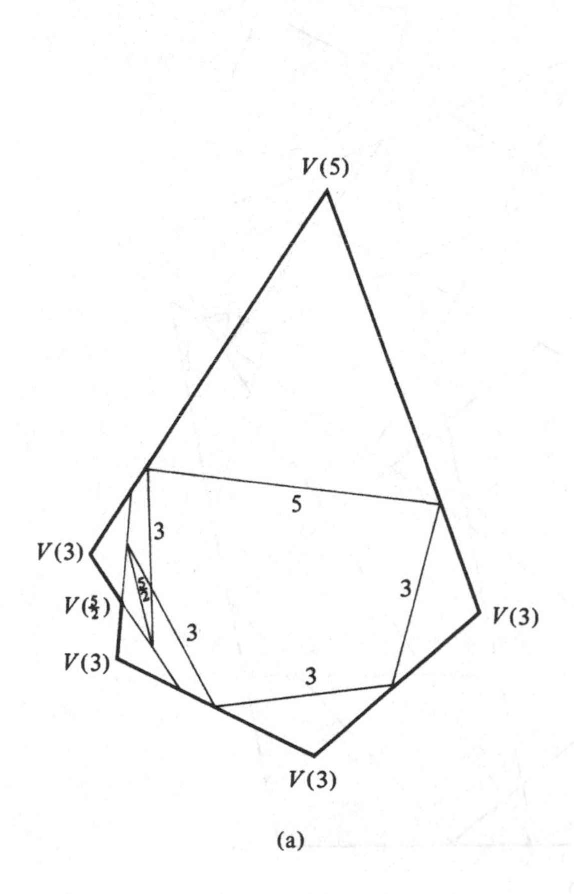

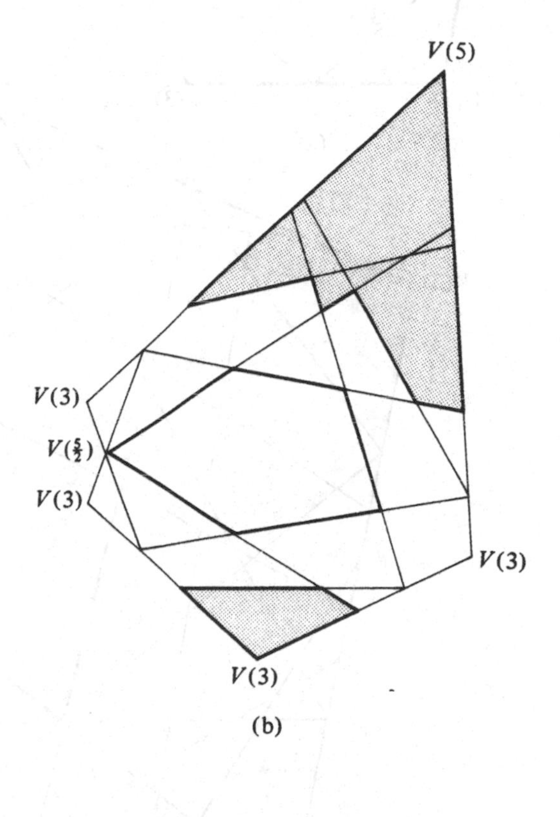

Fig. 101. Dorman Luke construction and stellation pattern for dual of **112**.

It is particularly noteworthy as a good example of an isohedral solid of a very complex shape with very simple facial planes, sixty semiregular congruent pentagons. So it is really a great pentagonal hexecontahedron. You might be wondering if the pentagons can be adjusted so as to become regular and thereby produce a new uniform polyhedron – a seventy-sixth to the set of seventy-five already known? The answer is that such an adjustment is indeed possible, but the result is a compound of five interpenetrating regular dodecahedra. By definition, a uniform polyhedron may not be a compound. The five dodecahedra will be presented in the final section of this book, where they turn up as a stellated form of a deltoidal hexecontahedron, not as a stellated form of a pentagonal hexecontahedron.

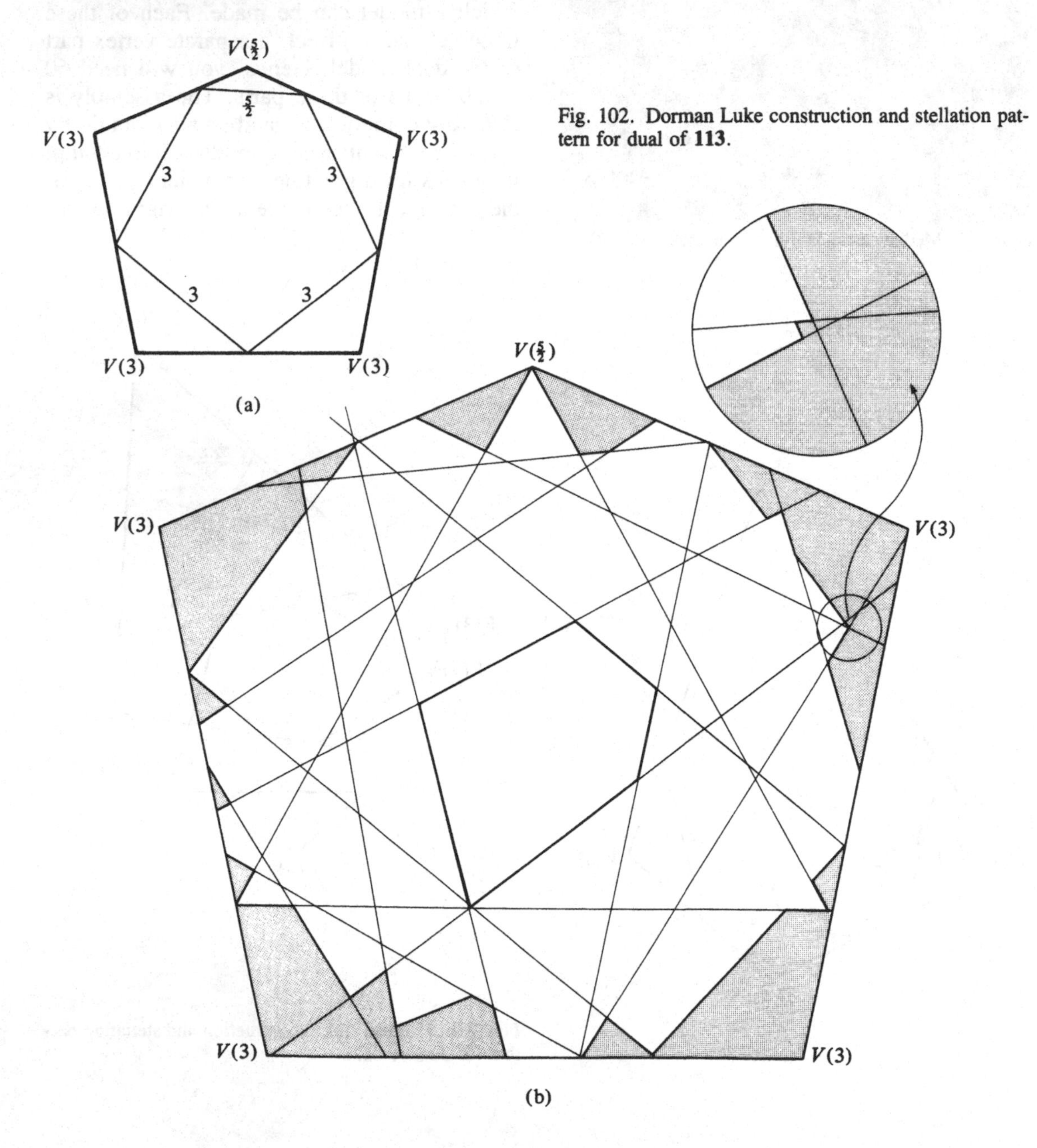

Fig. 102. Dorman Luke construction and stellation pattern for dual of **113**.

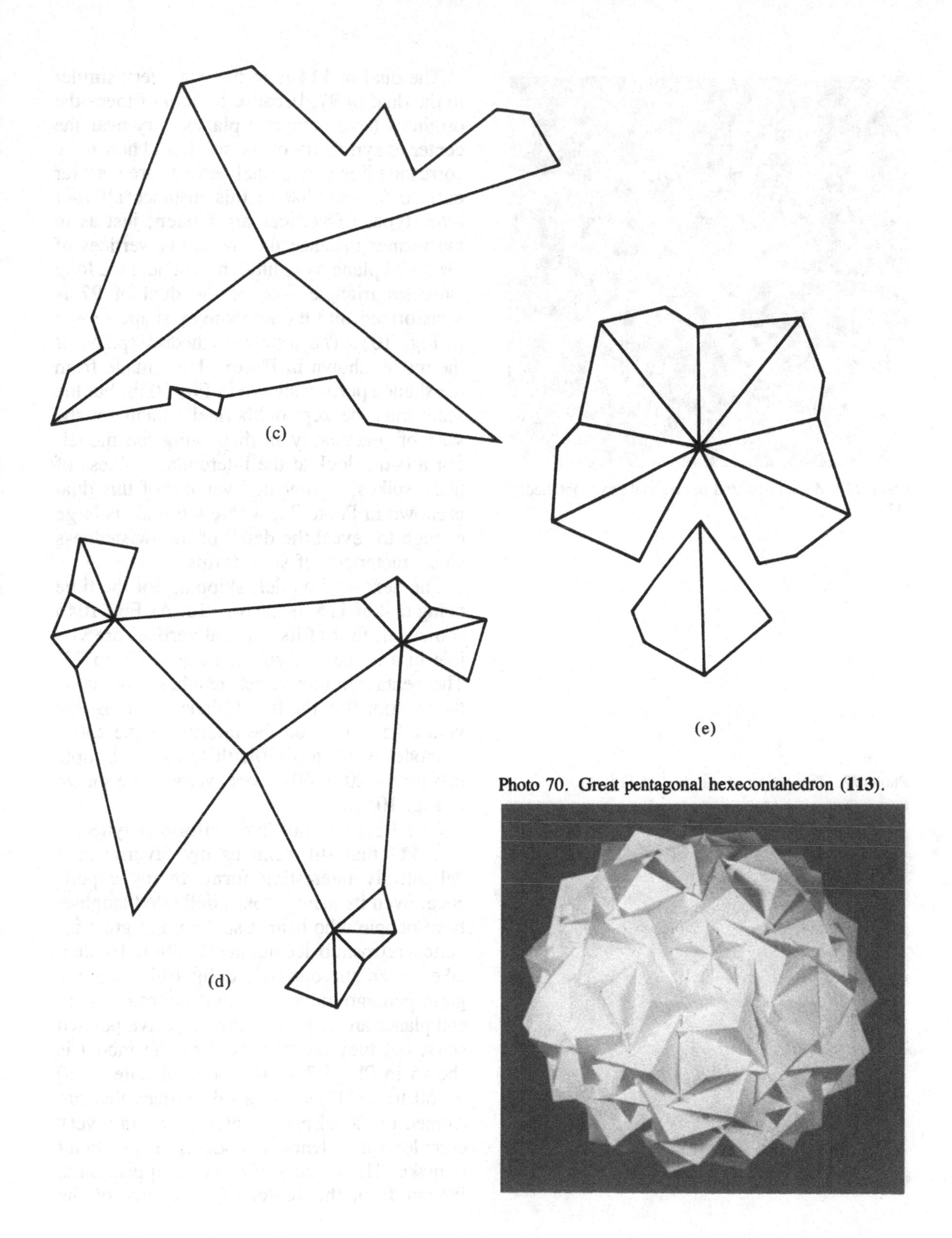

Photo 70. **Great pentagonal hexecontahedron (113).**

Photo 71. Medial inverted pentagonal hexecontahedron (**114**).

Photo 72. Truncation of the dual of **114**.

The dual of **114** is in one way very similar to the dual of **97**, because in both of these the originals have pentagon planes very near the center of symmetry of the solid, and hence the corresponding pentagonal vertices are very far out, so far out that in this instance all four other types of vertices are hidden, just as in the former instance the two other vertices of the facial plane were hidden. But here the long isosceles triangle face of the dual of **97** is transformed into the nonconvex shape shown in Fig. 103. The long pentahedral spikes of the model shown in Photo 71 are made from the shaded portion shown in Fig. 103b, but the scale must be kept deliberately small for the sake of practicality in displaying the model. For a better look at the interconnectedness of these spikes, a truncated version of this dual is shown in Photo 72, where the scale is large enough to reveal the detail of the twistedness so characteristic of snub forms.

The next dual model, skipping for the time being that of **115**, is that of **116**. As Fig. 104b shows, all four of its trigonal vertices are visible in a model, as you can see in Photo 73. The pentagrammic vertex reaches a point on the exterior, but the facial planes forming this vertex lie hidden on the interior of the solid. A model is not too difficult to make. Simple nets for the 20 + 60 trigonal vertices are shown in Fig. 104c.

Of all the seventy-five uniform polyhedra, it is **117** that still remains my favorite as a delightfully interesting form. In my experience, even its name evokes delighted laughter from people who hear it spoken: the great inverted retrosnub icosidodecahedron. Its dual takes on an appropriately delightful name: the great pentagrammic hexecontahedron. Its facial planes are sixty pentagrams or five-pointed stars, but they are not regular. The model is shown in Photo 74a. It is a composite of 20 + 60 trigonal spike-shaped vertices that are formed by facial planes intersecting in a very complex way. Hence, a model is very difficult to make. The vertices of this dual appear quite far out from the center of symmetry of the

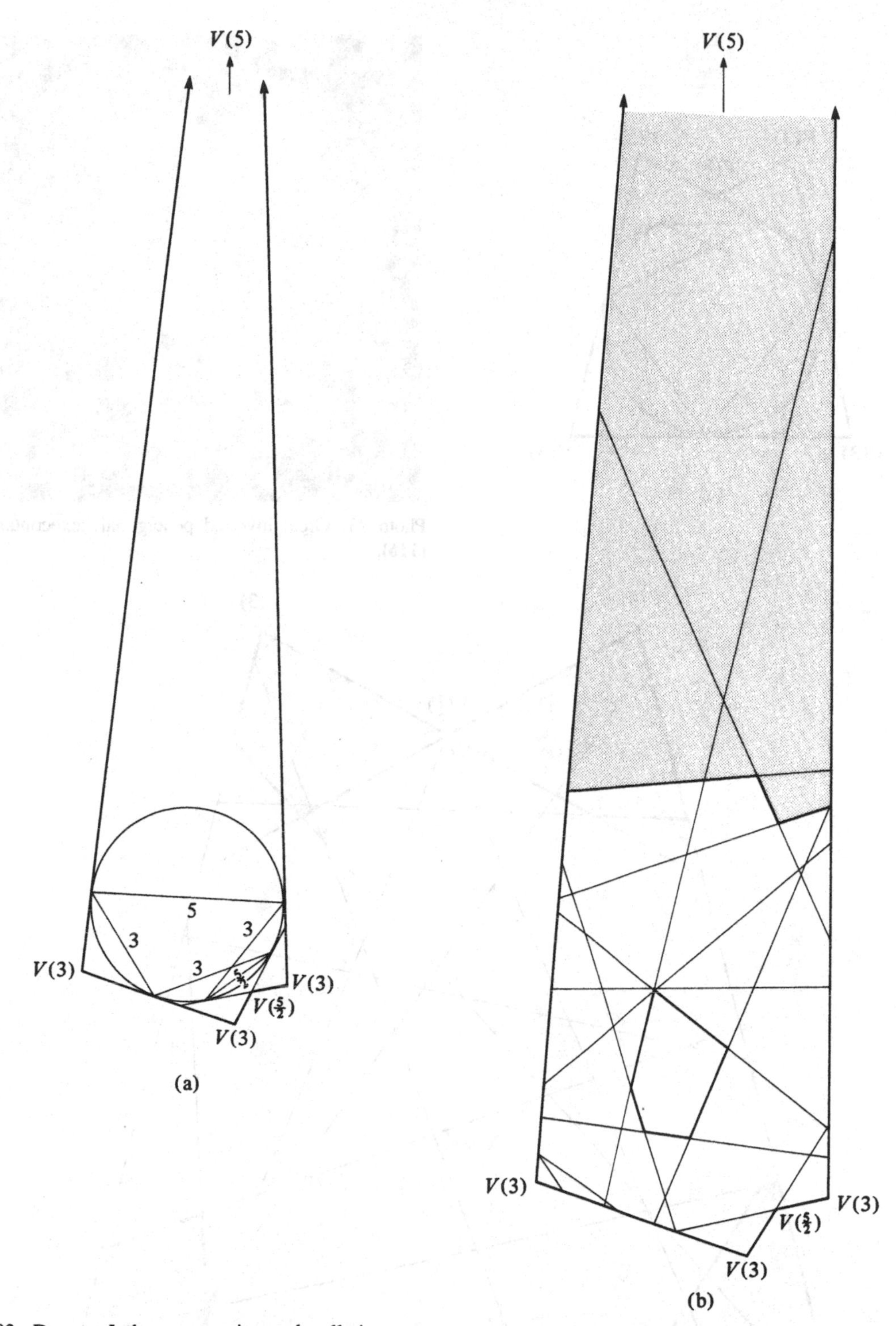

Fig. 103. Dorman Luke construction and stellation pattern for dual of **114**.

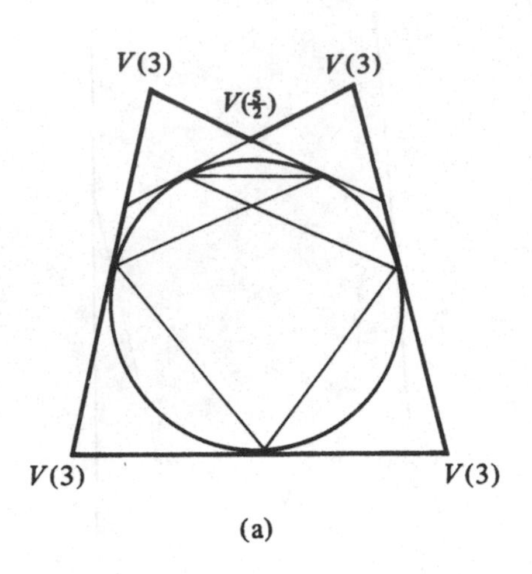

(a)

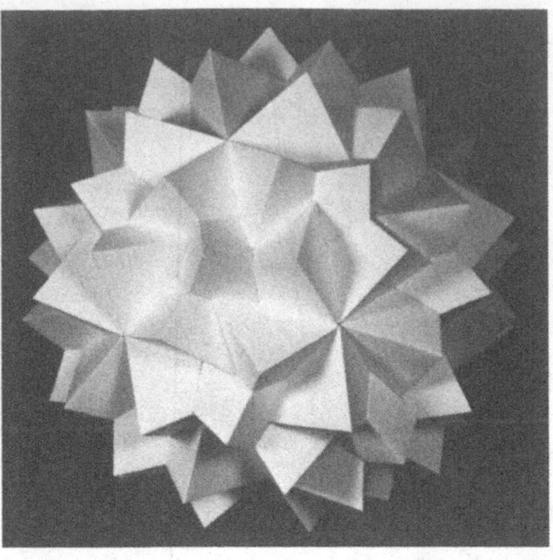

Photo 73. Great inverted pentagonal hexecontahedron (**116**).

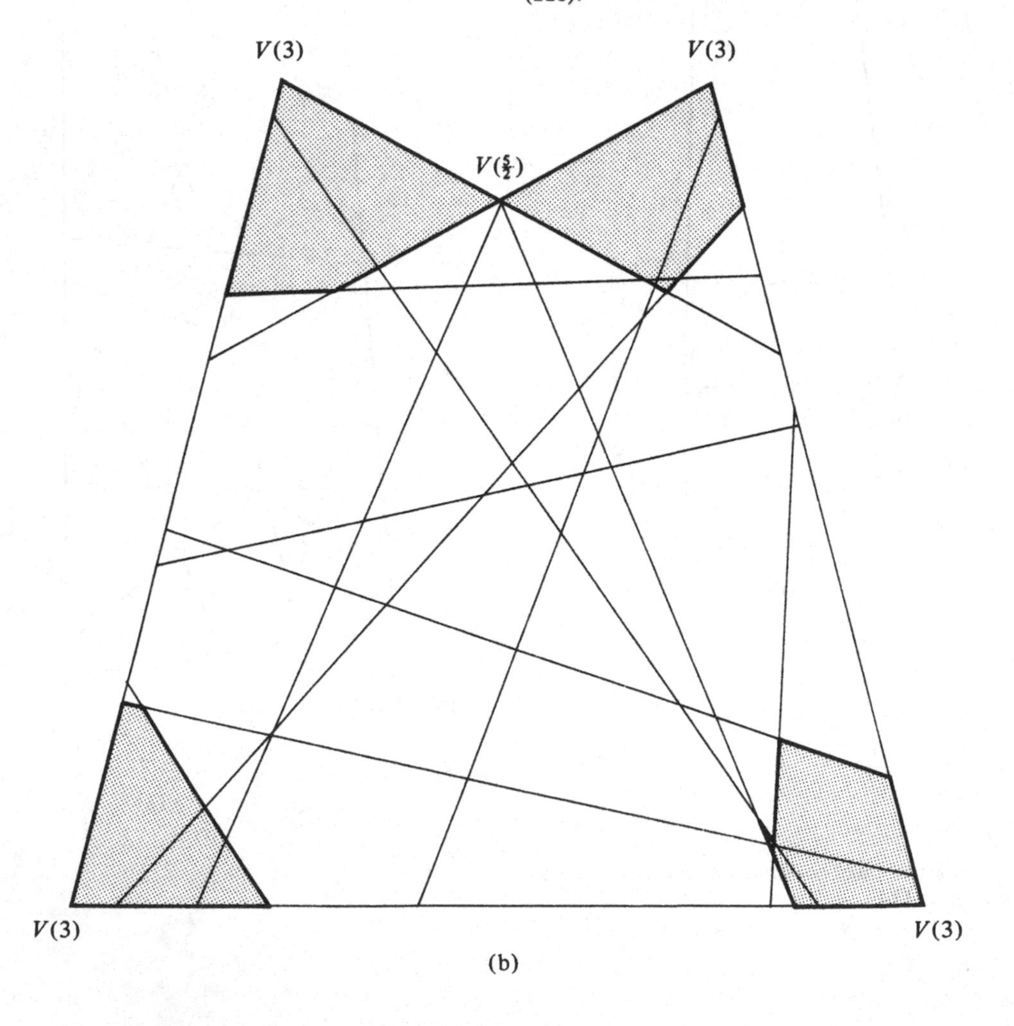

(b)

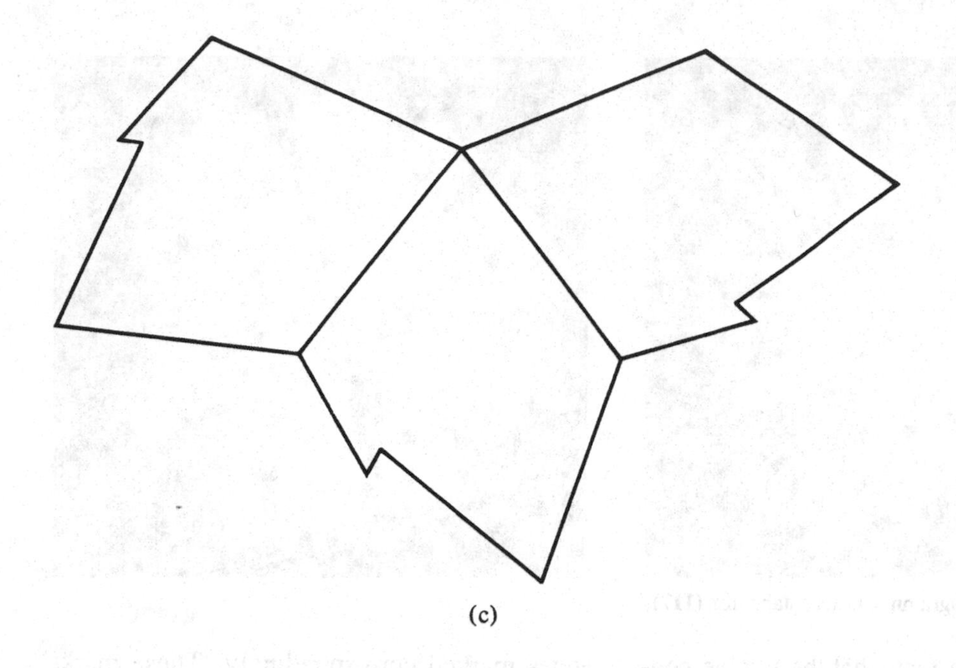

(c)

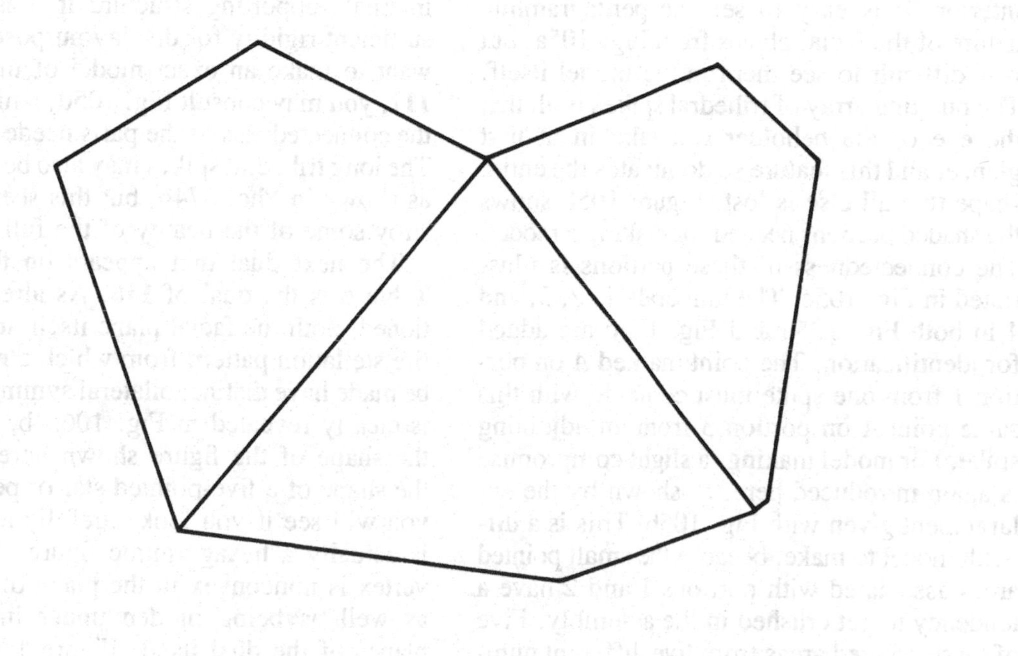

(d)

Fig. 104. Dorman Luke construction and stellation pattern for dual of 116.

Photo 74a,b. Great pentagrammic hexecontahedron (**117**).

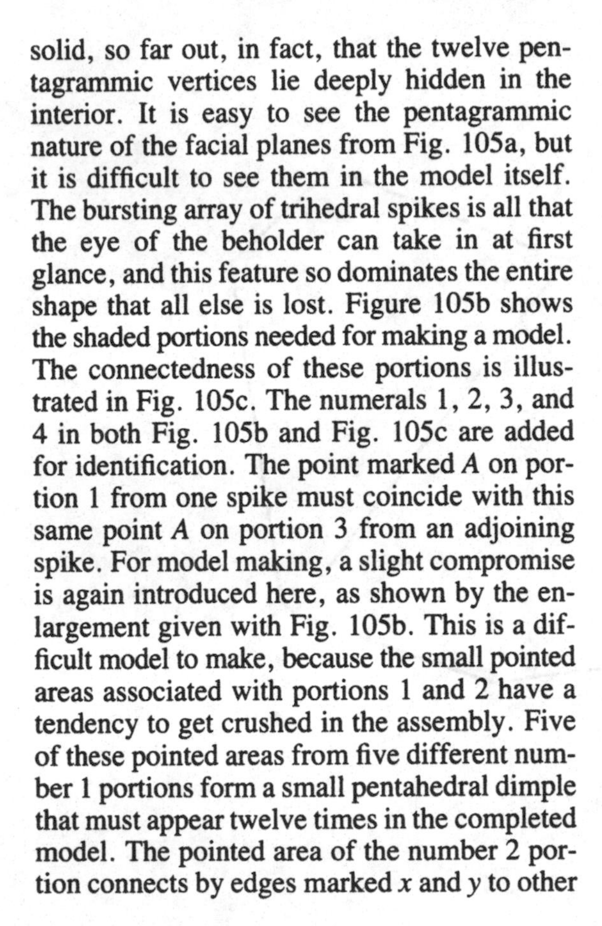

solid, so far out, in fact, that the twelve pentagrammic vertices lie deeply hidden in the interior. It is easy to see the pentagrammic nature of the facial planes from Fig. 105a, but it is difficult to see them in the model itself. The bursting array of trihedral spikes is all that the eye of the beholder can take in at first glance, and this feature so dominates the entire shape that all else is lost. Figure 105b shows the shaded portions needed for making a model. The connectedness of these portions is illustrated in Fig. 105c. The numerals 1, 2, 3, and 4 in both Fig. 105b and Fig. 105c are added for identification. The point marked *A* on portion 1 from one spike must coincide with this same point *A* on portion 3 from an adjoining spike. For model making, a slight compromise is again introduced here, as shown by the enlargement given with Fig. 105b. This is a difficult model to make, because the small pointed areas associated with portions 1 and 2 have a tendency to get crushed in the assembly. Five of these pointed areas from five different number 1 portions form a small pentahedral dimple that must appear twelve times in the completed model. The pointed area of the number 2 portion connects by edges marked x and y to other edges marked correspondingly. These markings are shown in Fig. 105c. The model needs internal supporting structure if it is to have sufficient rigidity for display purposes. If you want to make an exact model of the dual of **117**, you may consult Fig. 105d, which shows the connectedness of the parts needed for this. The long trihedral spikes may also be truncated as shown in Photo 74b, but this seems to destroy some of the beauty of the full model.

The next dual that appears on the list in Table 6 is the dual of **118**. As already mentioned, both the facial plane itself and the entire stellation pattern from which a model can be made have distinct bilateral symmetry. This is clearly revealed in Fig. 106a–b. Although the shape of the figure shown here suggests the shape of a five-pointed star or pentagram, you will see if you look carefully at it that it is actually a hexagrammic figure. The $V(5/2)$ vertex is nonconvex in the plane of the face, as well as being hidden under intersecting planes of the dual itself. Figure 106b shows the shaded portions from which a model can be made. The connectedness of these is shown in the nets laid out in Fig. 106c. Photo 75 shows that an extremely attractive model can

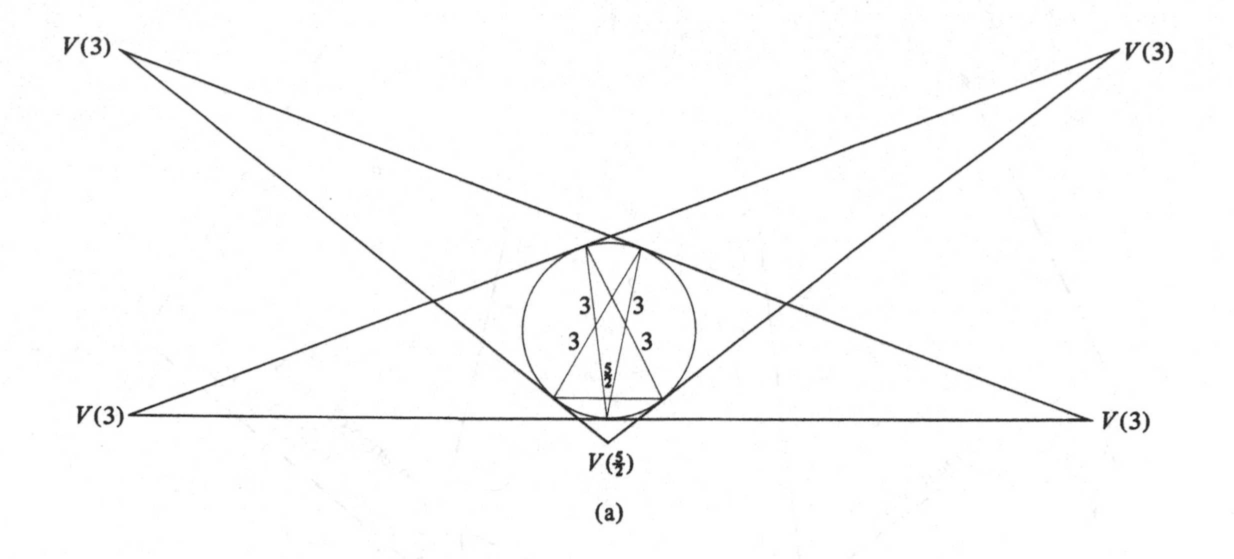

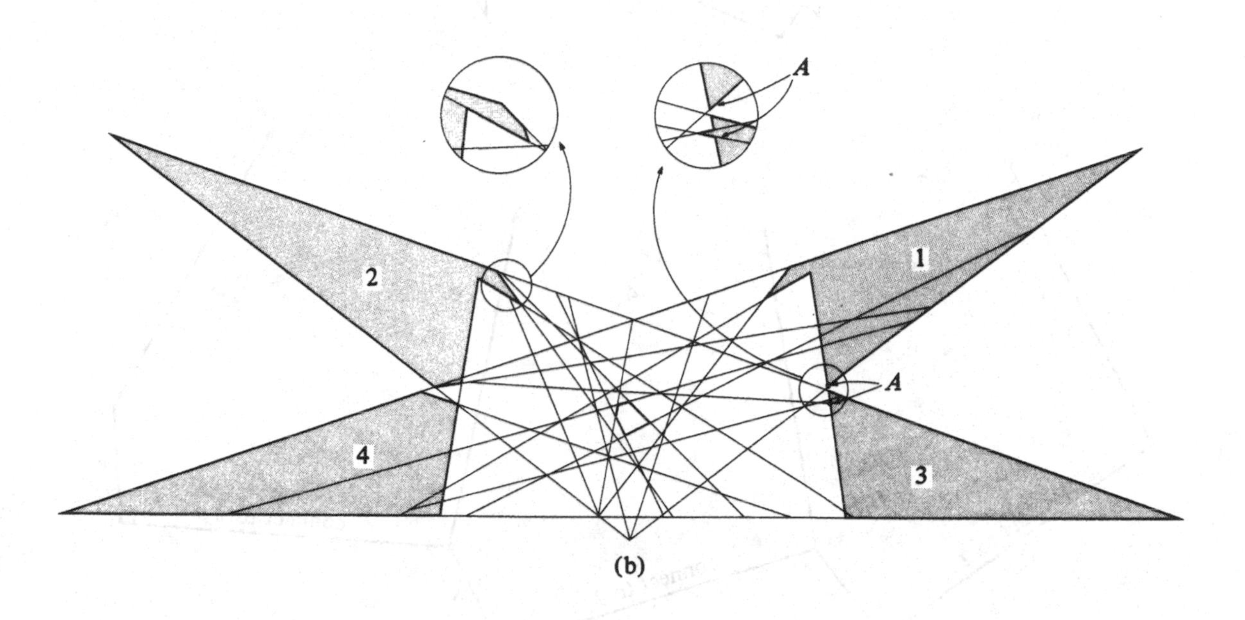

Fig. 105. Dorman Luke construction and stellation pattern for dual of **117**.

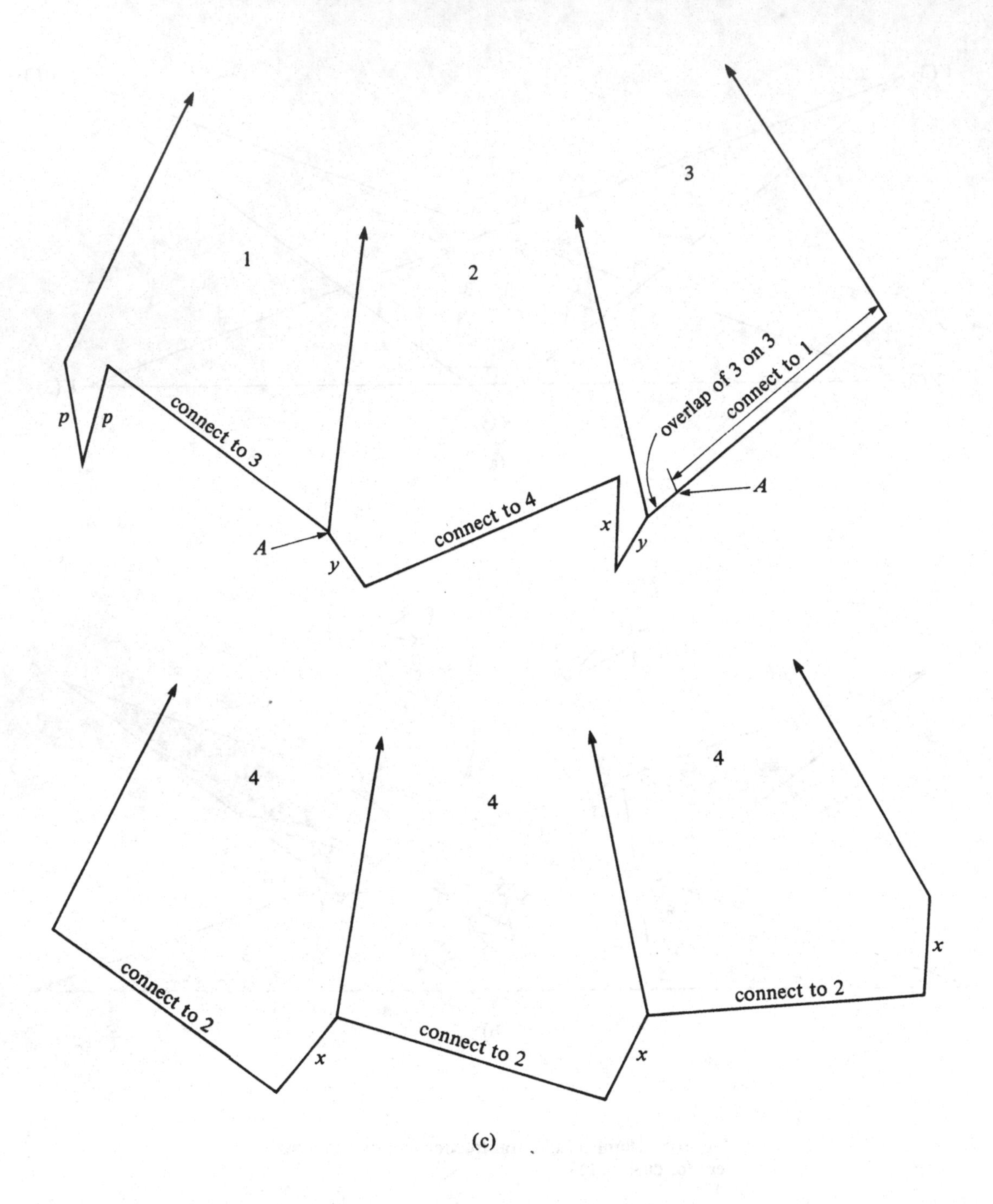

(c)

Fig. 105 (*cont'd.*)

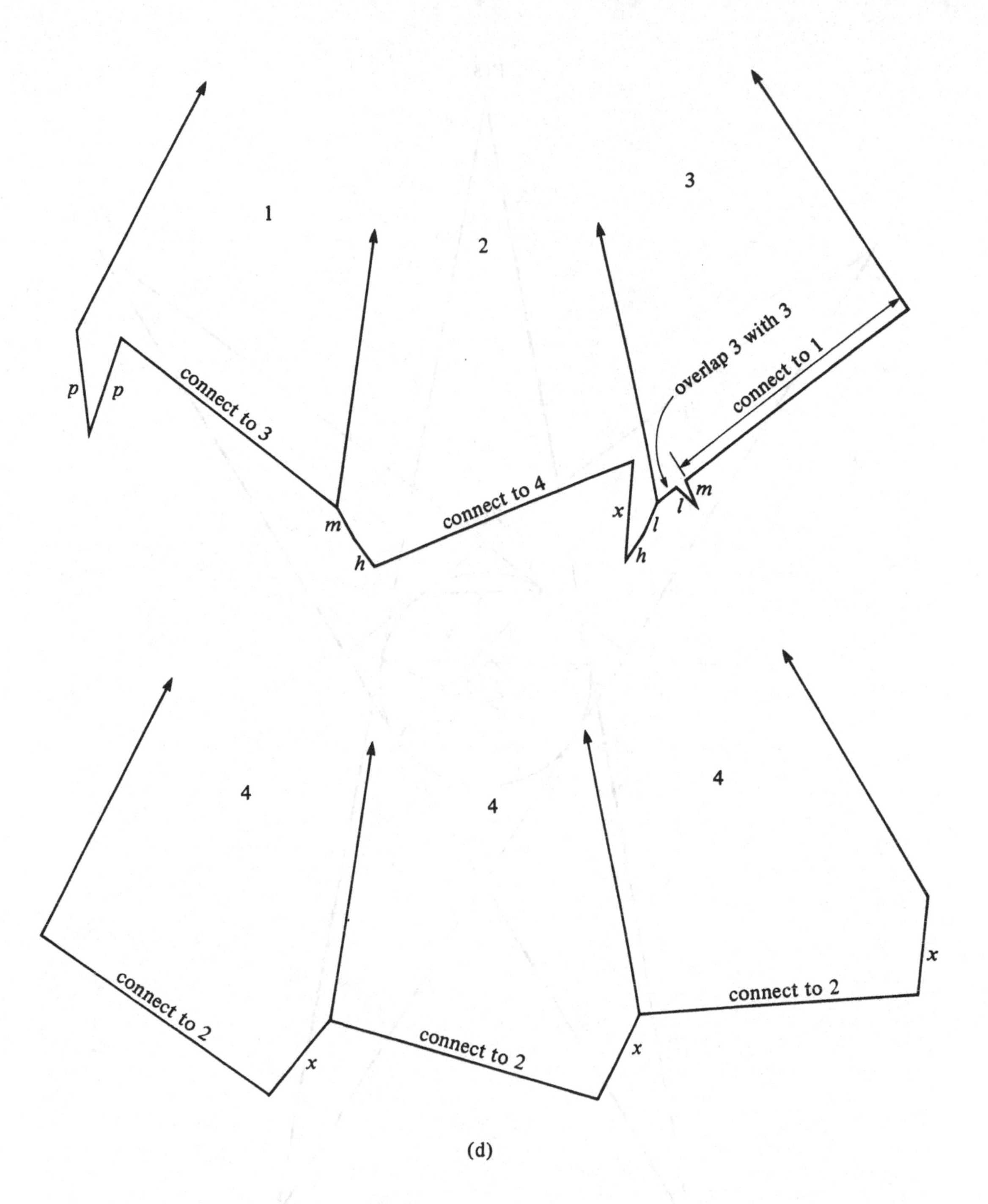
1
2
3
p
p
connect to 3
m
h
connect to 4
x
h
l
l
m
overlap 3 with 3
connect to 1
4
4
4
connect to 2
x
connect to 2
x
connect to 2
x

(d)

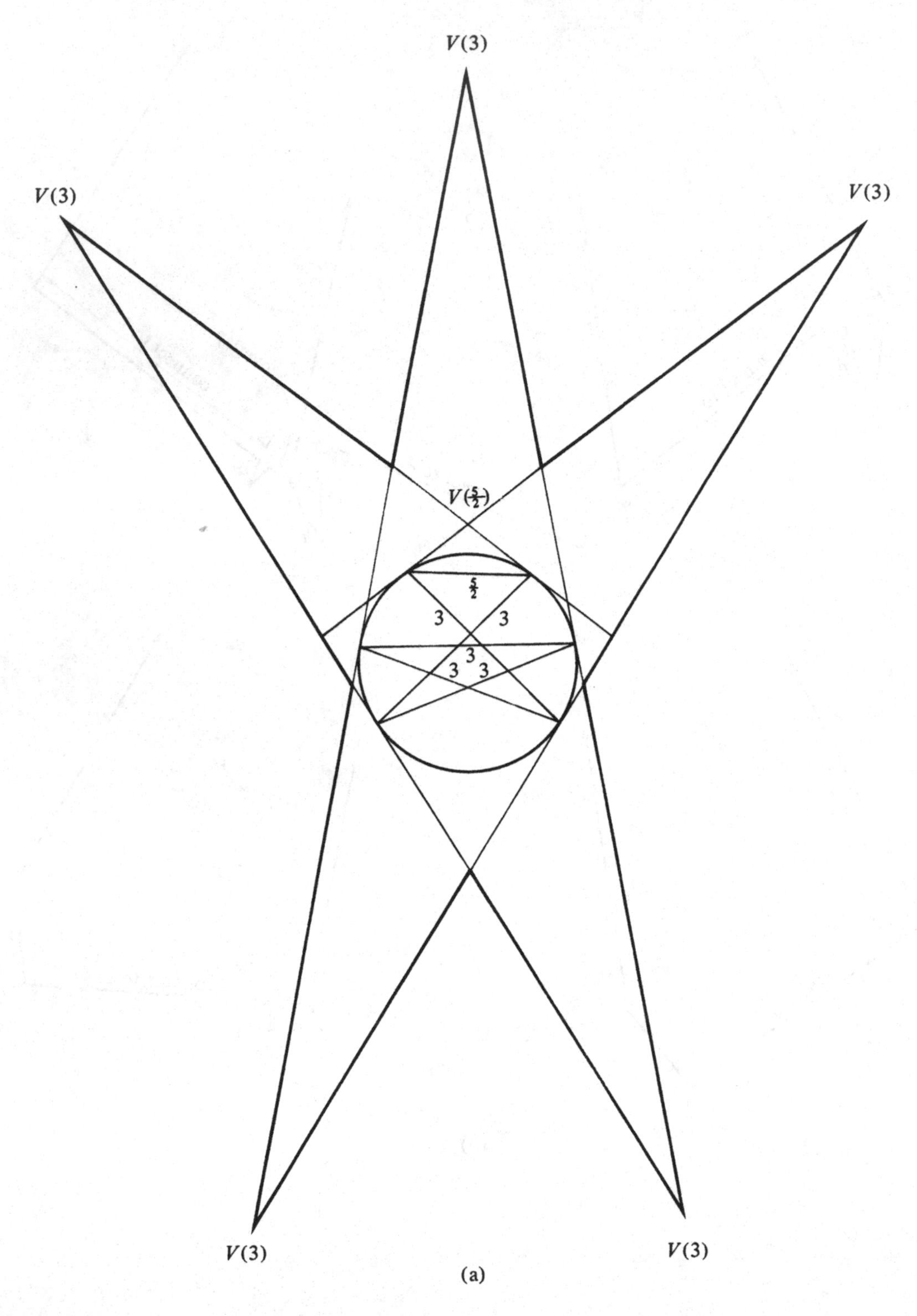

(a)

Fig. 106. Dorman Luke construction and stellation pattern for dual of **118**.

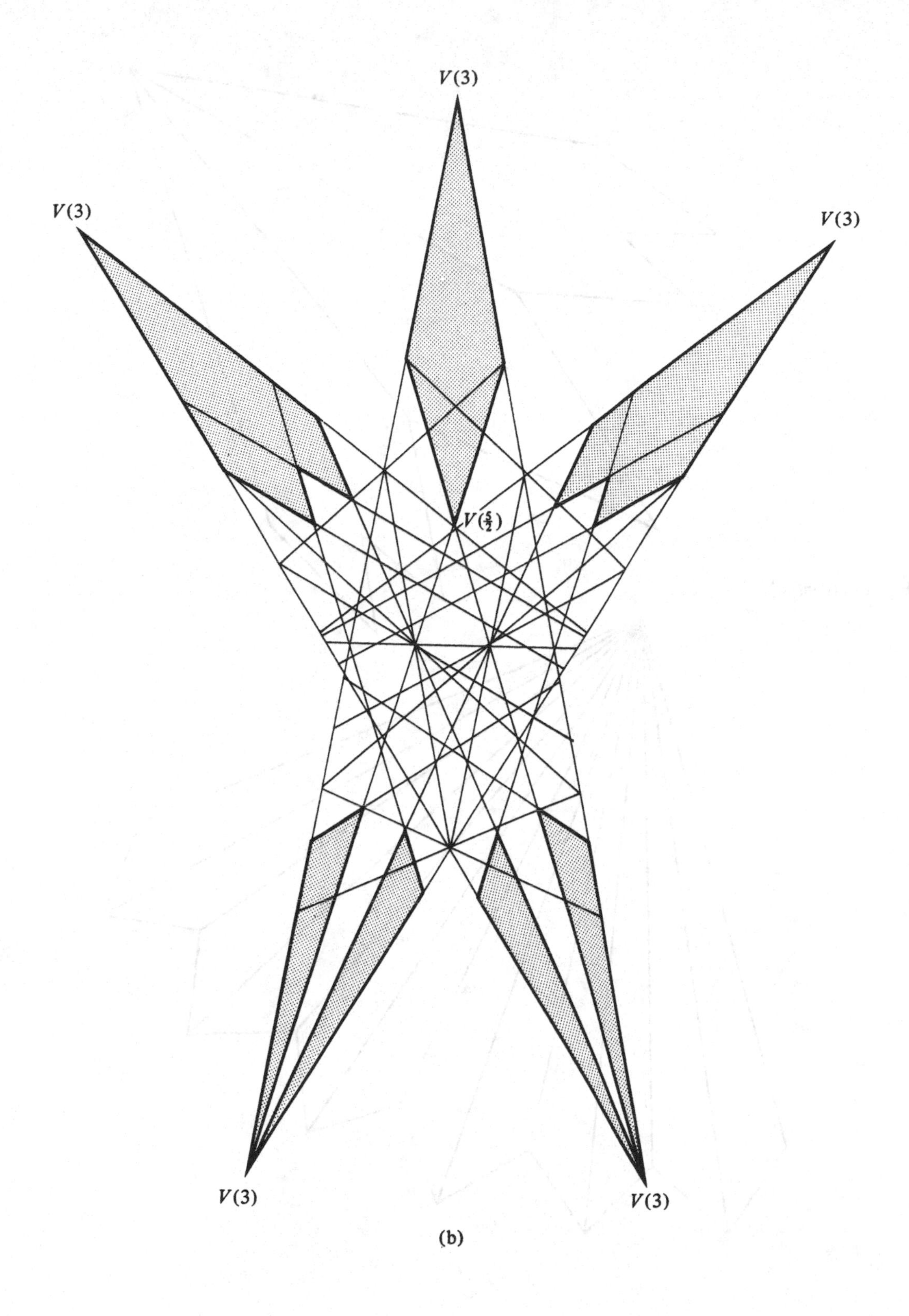

(b)

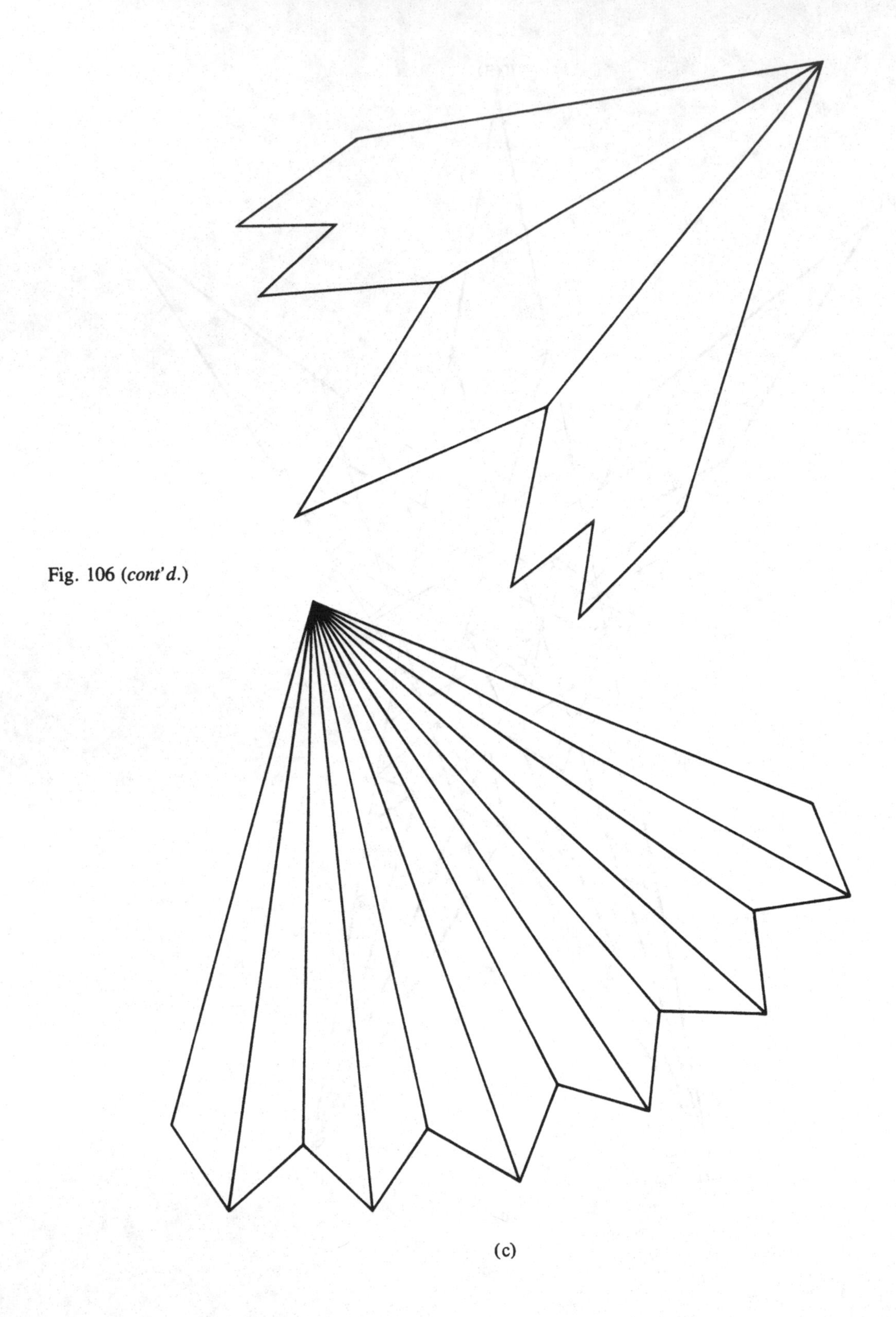

Fig. 106 (*cont'd.*)

(c)

Photo 75. Small hexagrammic hexecontahedron (**118**).

be made from the drawings shown here. You must have a great deal of expertise in model making to get all the parts together, but if you have been making models you know that skill comes from prolonged experience. Once you possess this, you will find that this model is not as difficult to assemble as it may look. However, great care is needed in cementing the dodecagrammic vertex parts or spikes (the double triangular vertices) into hexahedral dimples to get the twistedness of these spikes properly oriented. When completed, this makes a very rigid as well as attractive model.

The dual of **115** is shown in Photo 76, a rather attractive model in its own twisted way, having 20 + 60 trigonal vertices, each with lateral surfaces whose edges meet at right angles to each other. This reveals itself clearly in the stellation pattern shown in Fig. 107b, where the core of the stellation process is a deltoidal hexecontahedron. Two separate nets, one for each type of vertex, will be needed for the model. These are easily made from the shaded portions shown, and the assembly is easily detected from the photo. The relationship that this dual has with **119** is manifested by the fact that both are derived from the stellation process applied to the same core. Moreover, the beginnings of the octagrammic or double-square facial planes that belong to the dual of **119** are here already detected as beginning to take shape.

The last of the complete set of duals for the seventy-five uniform polyhedra, the dual of **119**, will now be considered. The original **119** itself is really a hemipolyhedron, because all of the sixty double-square facial planes pass through the center of symmetry of the solid. Its dual, therefore, must have corresponding vertices at infinity. In this sense it is like all the other hemipolyhedra listed in Table 5, but it is also something like the other nonconvex snubs as well. Hence, it is listed here at the end of Table 6. The dual of **119** is shown in Photo 77. As you can immediately see, it has the same characteristic appearance that the duals of the other hemipolyhedra display. It may simply be described as a compound of sixty (or, more correctly, thirty) double-square or octagrammic prisms whose centers of symmetry radiate out along the same axes of symmetry as those belonging to a rhombicosidodecahedron. Hence, the core of the stellation pattern is a deltoidal hexecontahedron, as you can see from Fig. 108b. The strange polygon that actually should be its facial plane is shown in Fig. 108c, where four vertices, the even-numbered ones, are at infinity, and the odd-numbered vertices are in finite space. The shaded portions of Fig. 108b show the parts needed for making a model. Because the connectedness of these portions is very complex, they are numbered, and the same numbering identifies these in Fig. 108e, which is the drawing for a net to be used in making one of the octagrammic prisms. The octagram that is the truncating plane is shown in Fig. 108d. Note that it is not a regular octagram. The squares do not have a 45-degree twist relative to each other. A little experimentation should reveal to you how the jagged edges meet to connect one prism to another until all sixty are assembled. This model, like the other stellations to infinity, really has the appearance of an explosion in progress. In fact,

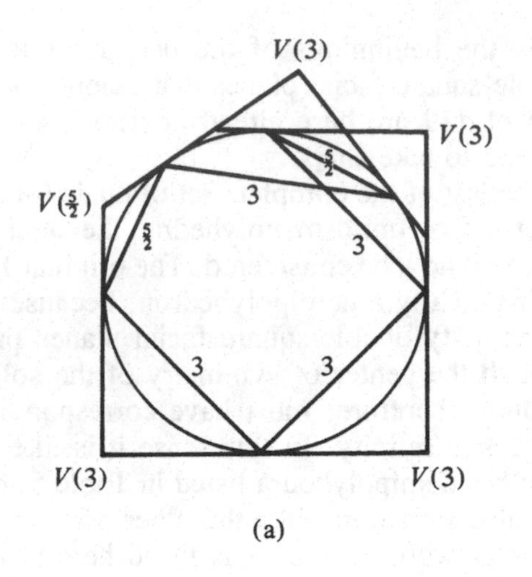

Photo 76. Great hexagonal hexecontahedron (**115**).

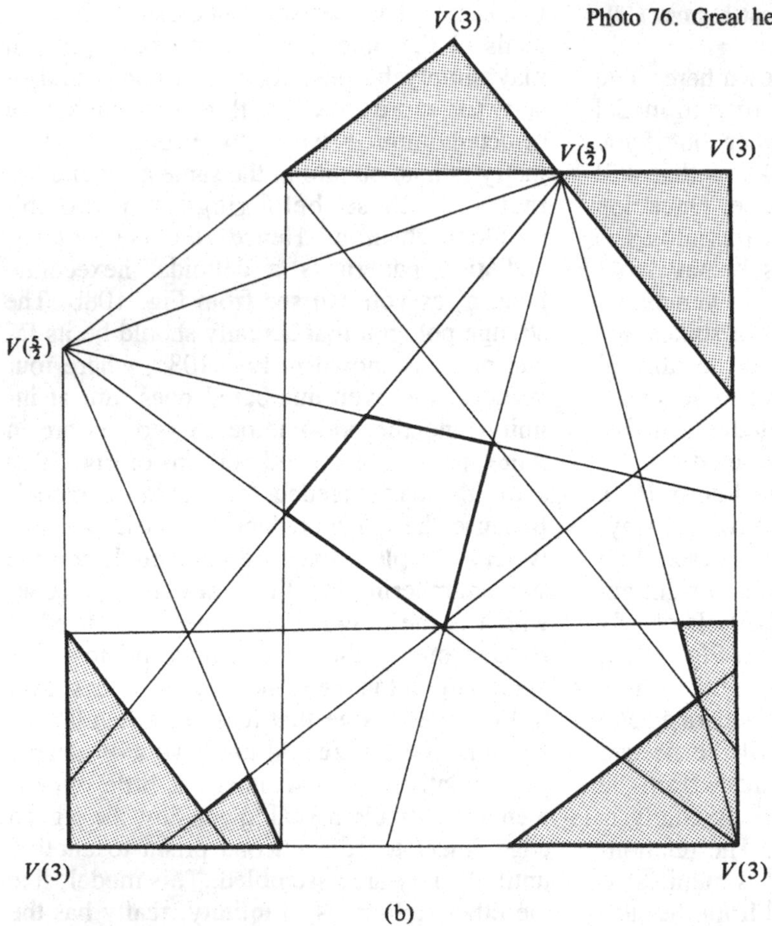

Fig. 107. Dorman Luke construction and stellation pattern for dual of **115**.

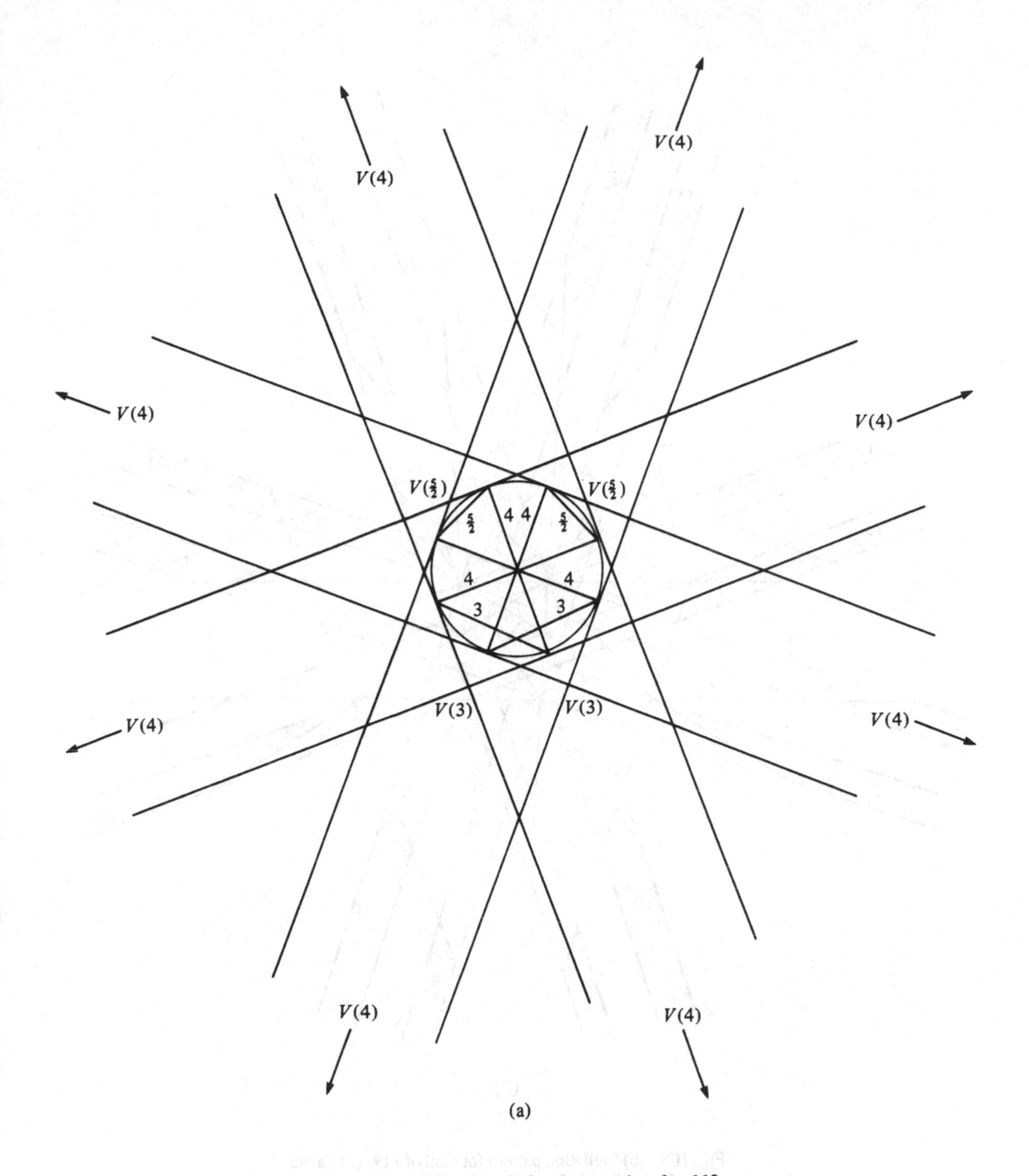

Fig. 108. (a) Dorman Luke construction for **119**.

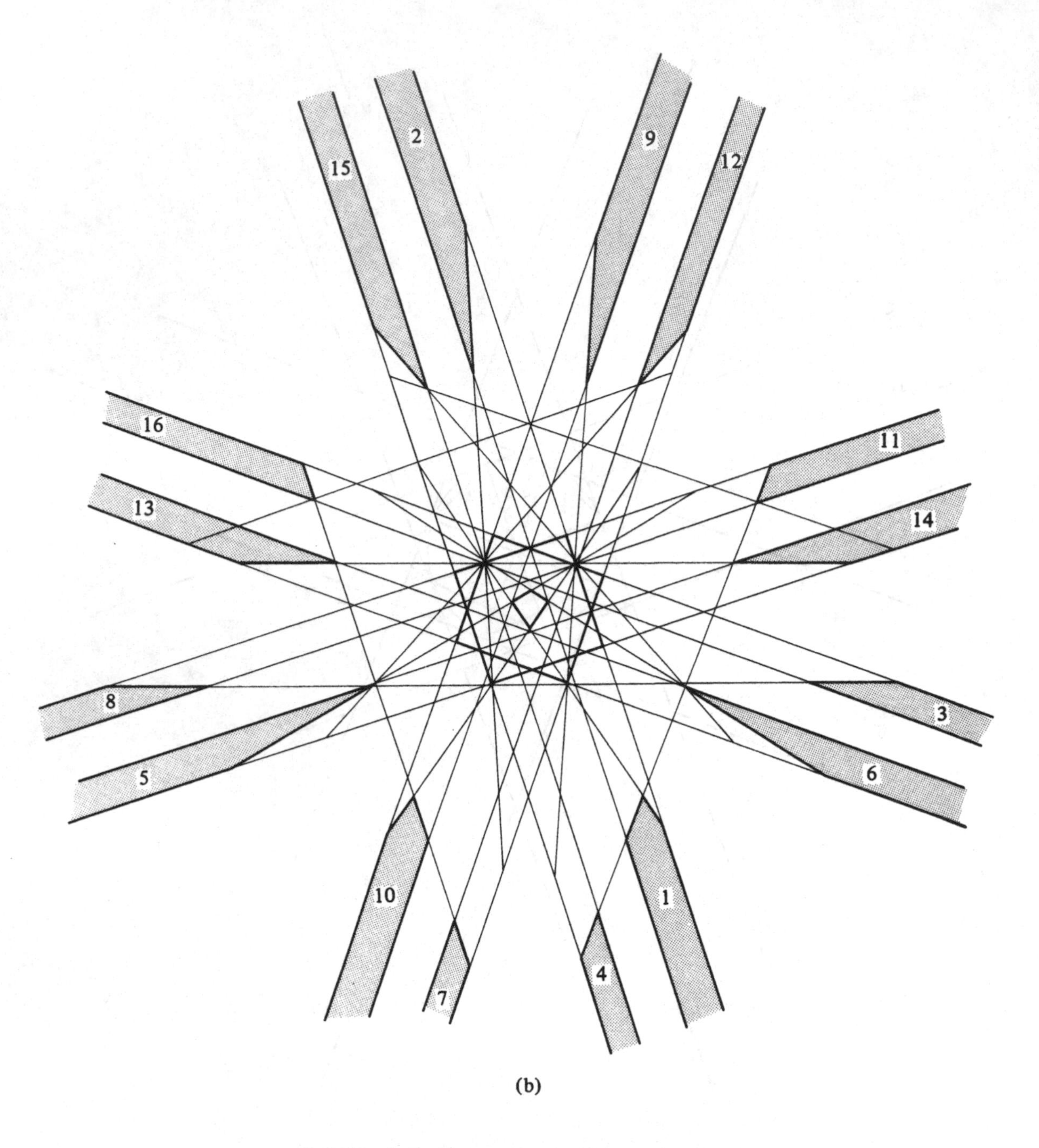

Fig. 108. (b) Stellation pattern for dual of **119**. (c) Facial plane for dual of **119**, with vertices at infinity.

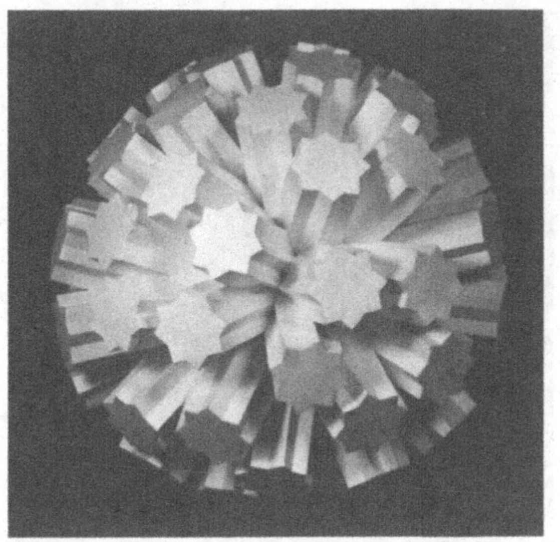

Photo 77. Great dirhombicosidodecacron (**119**).

it easily suggests how all the octagrammic facial planes began their journey outward from the position they originally occupied in the uniform polyhedron **119** itself. Then, leaving the center of symmetry but confining their journey as facial planes normal to the axes of symmetry of a rhombicosidodecahedron, these facial planes reach an intermediary stage at which they become the facial planes of a compound of twenty cubes. A model of this is shown in Photo 78. Figure 108f gives the stellation pattern for this, and the shaded portions here are connected as shown in Fig. 108g, illustrating the nets used to make this model.

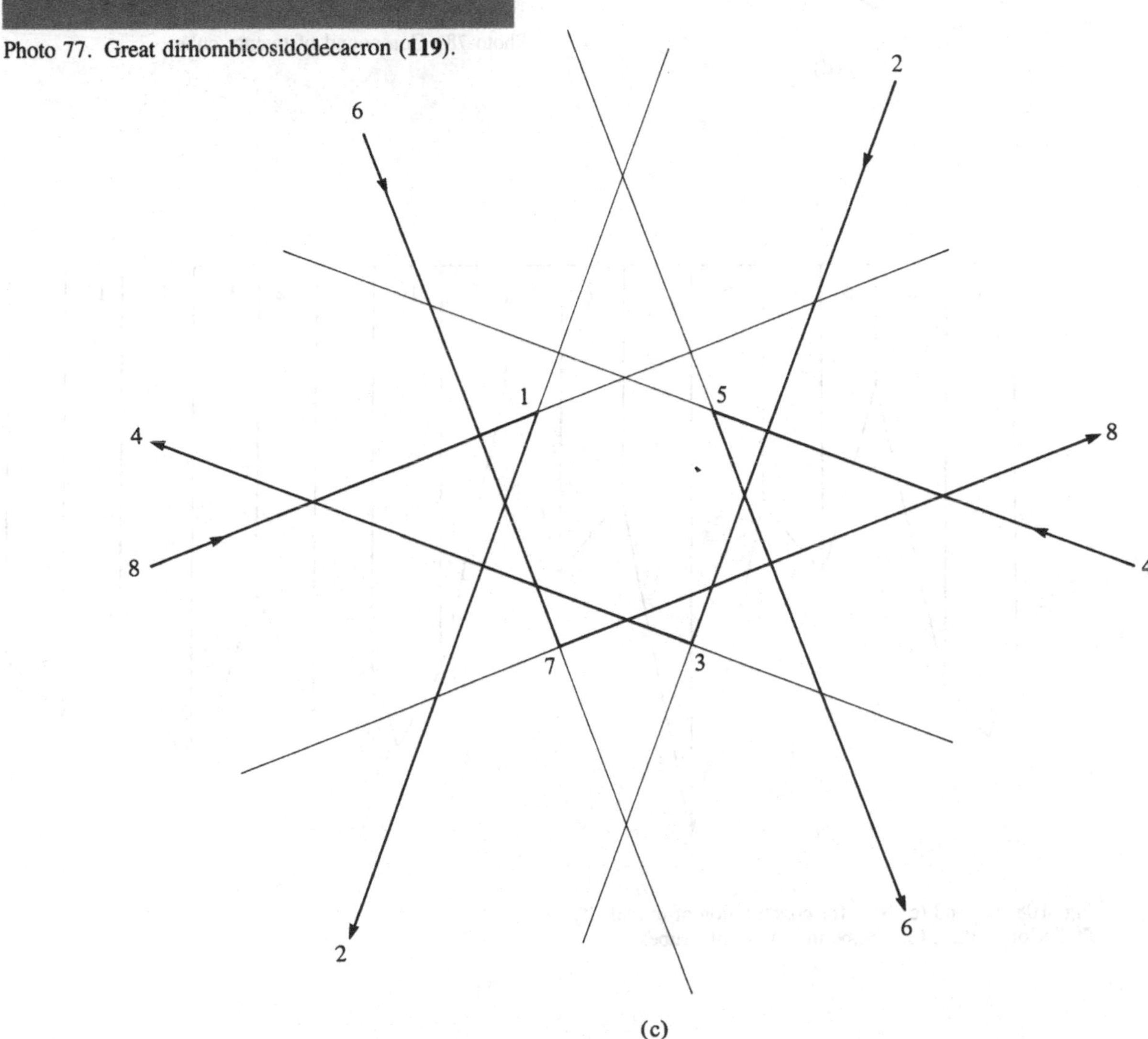

(c)

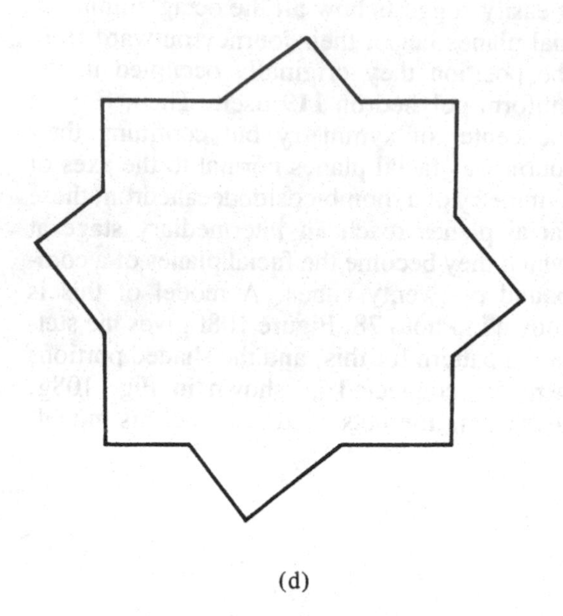

(d)

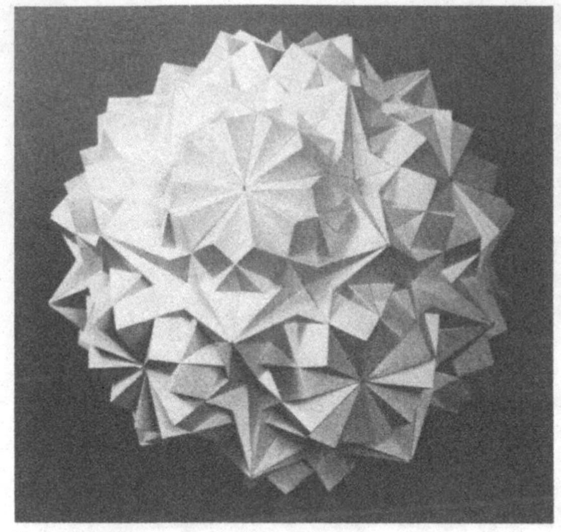

Photo 78. Compound of twenty cubes.

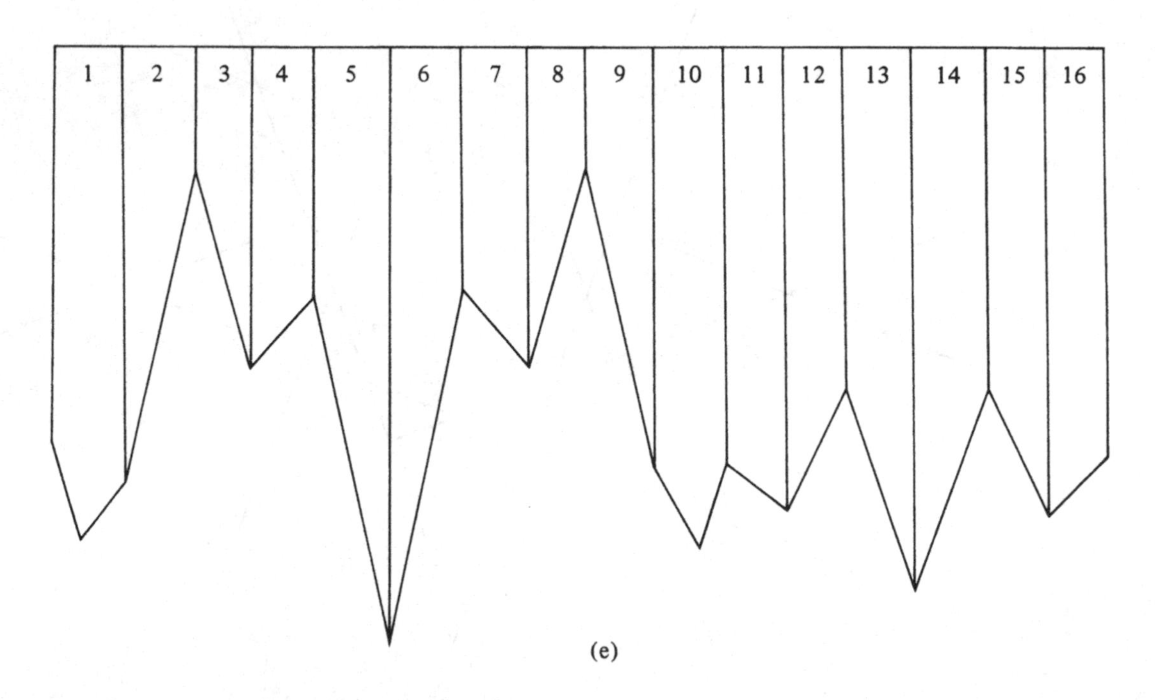

(e)

Fig. 108. (d) and (e) Nets for construction of model. (f) Stellation pattern for compound of twenty cubes.

Finally, the octagrammic facial planes continue their explosive journey into three-dimensional space, to be caught or frozen into a model that must serve as the dual from which it was born. This model, which must be taken as a dual of **119**, is one finite stage in the journey of the octagrammic facial planes toward infinity. If in two-dimensional projective geometry it is sometimes helpful to the imagination to use the dictum that parallel lines meet at infinity, here it may be helpful to the imagination to think of the entire pencil of lines forming the edges of the octagrammic prisms as eventually meeting at infinity, at points that are the octagrammic vertices of the dual of **119**. This is truly a remarkable and awe-inspiring model, an appropriate ending for this presentation of duality for nonconvex uniform polyhedra.

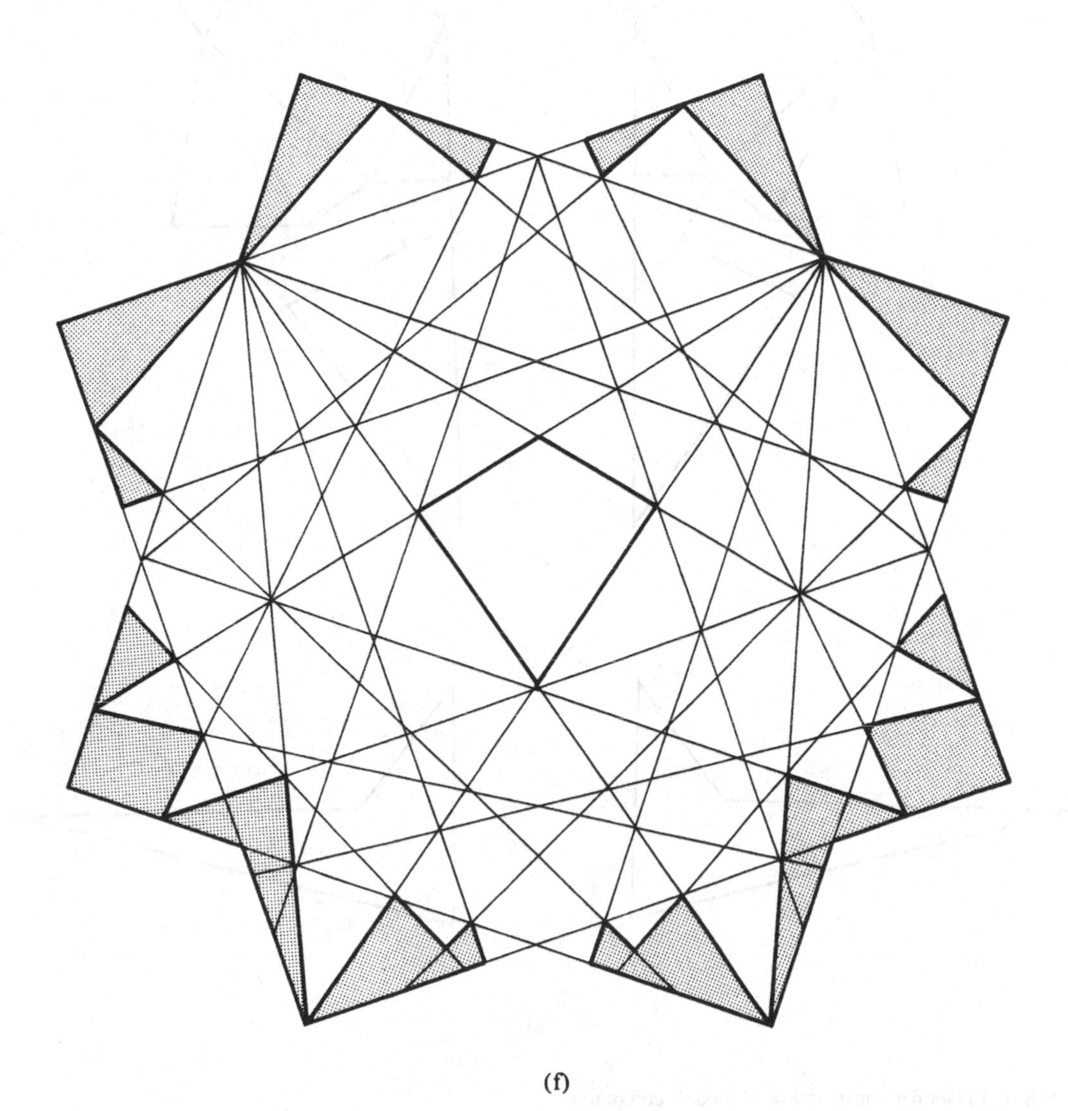

(f)

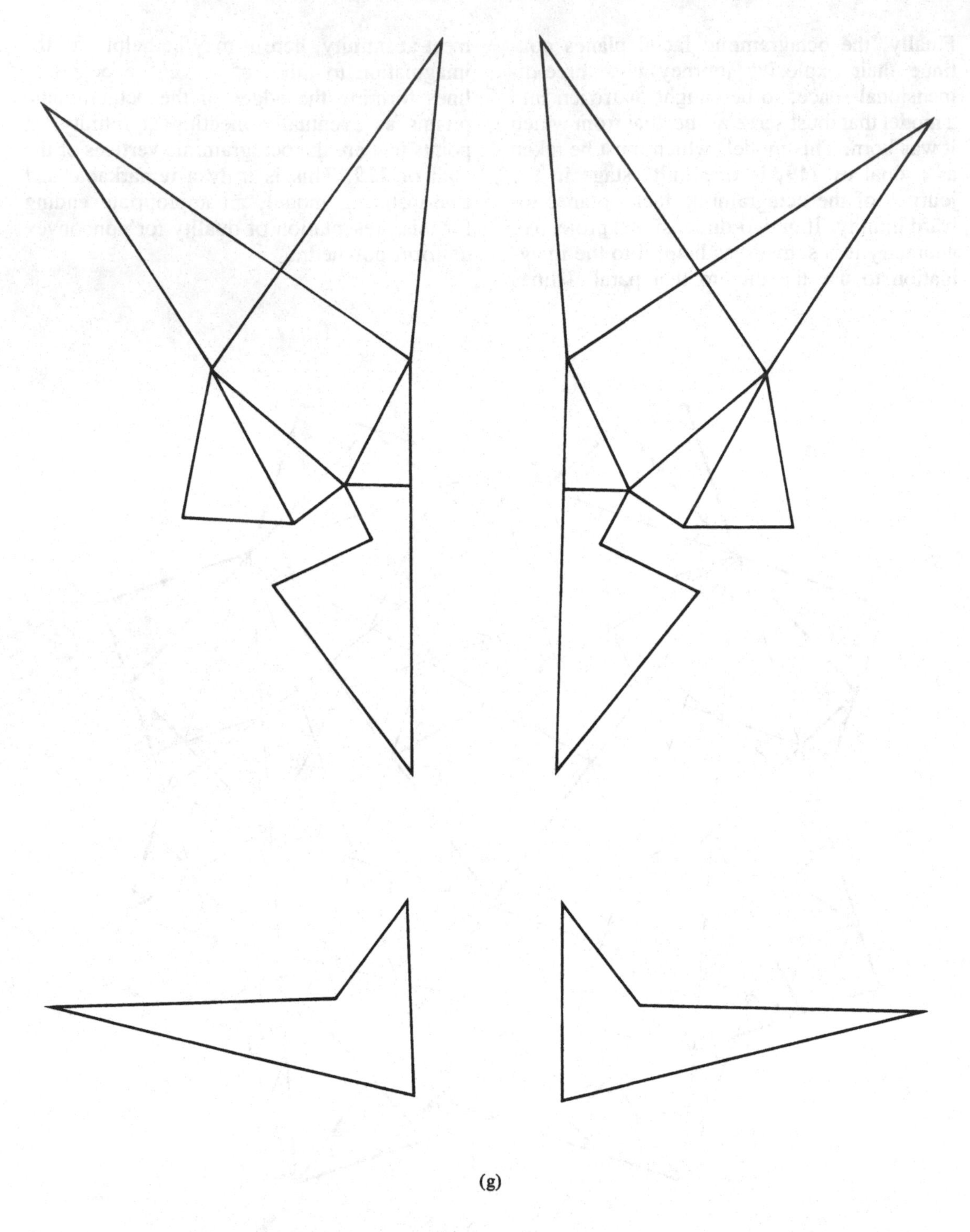

(g)

Fig. 108. (g) Nets for construction of model: compound of twenty cubes.

V. Some interesting polyhedral compounds

The fact that variations of Archimedean forms can lead to interesting results is well borne out by compounds of regular dodecahedra. An investigation of some of these forms will be the topic of this section.

Figure 109 shows the stellation pattern for compounds of two dodecahedra. Oddly enough, this pattern actually belongs to the octahedral symmetry group. Once you make the models shown in Photos 79 and 80, you

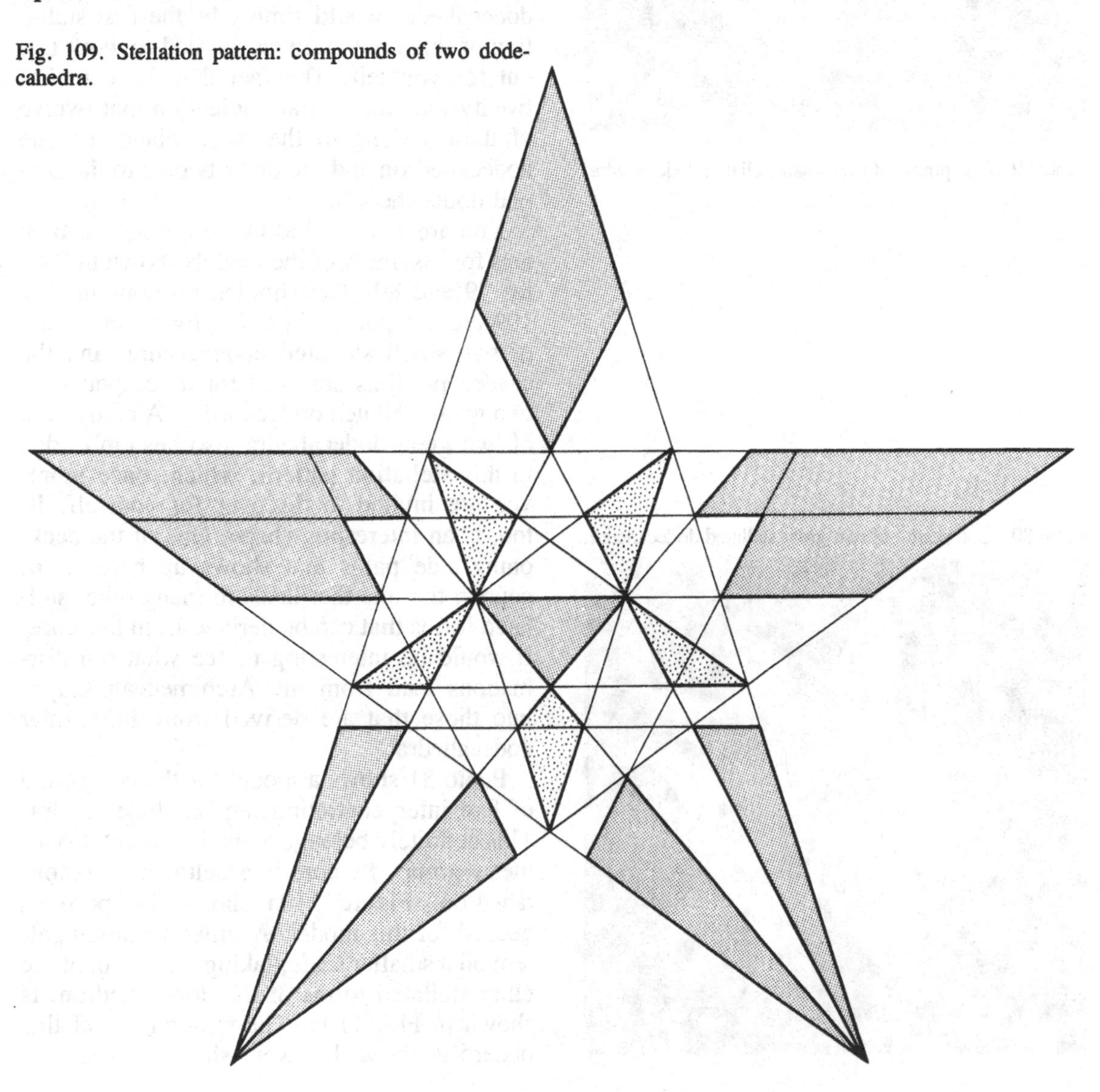

Fig. 109. Stellation pattern: compounds of two dodecahedra.

Photo 79. Compound of two small stellated dodecahedra.

Photo 80. Compound of two great stellated dodecahedra.

will easily see this for yourself. The core of this stellation is a tetrakishexahedron whose facial planes are made up of twenty-four isosceles triangles, nearly like those of the Archimedean form that is the dual of the truncated octahedron. From Fig. 109 you can easily find for yourself the angles for one of these triangles, because it is clearly embedded in a regular pentagon having its base made up of a diagonal of the pentagon and its vertex point on the midpoint of a side of this pentagon. The compound of two interpenetrating regular dodecahedra would simply be the first stellation of this core. You are invited to work this out for yourself. The fact that the core has twenty-four faces is an indication that twelve of these belong to the facial planes of one dodecahedron and the other twelve to the second dodecahedron.

You are also invited to work out your own nets for assembly of the models shown in Photos 79 and 80. The stippled portions in Fig. 109 are the portions needed for a compound of two small stellated dodecahedra, and the shaded portions are used for a compound of two great stellated dodecahedra. A compound of two great dodecahedra also lies embedded in this stellation pattern, which, once more, you are invited to discover for yourself. It, too, is an interesting shape. One of the beckoning side paths that shows up here is, of course, the one that leads to many other stellated forms that can be derived from this core. It would be interesting to see what transformations lead from the Archimedean shapes into those that are derived from the regular dodecahedra.

Photo 81 shows a model for the compound of five interpenetrating regular dodecahedra. This definitely belongs to the icosahedral symmetry group. Its core is a deltoidal hexecontahedron. Figure 110a shows the portions needed for this model. A fuller stellation pattern on a smaller scale, taking into account the other stellated forms of the dodecahedron, is shown in Fig. 110b. The relationship of this pattern to those that were shown in Figs. 58

Photo 81. Compound of five dodecahedra.

Fig. 110. (a) Compound of five regular dodecahedra.

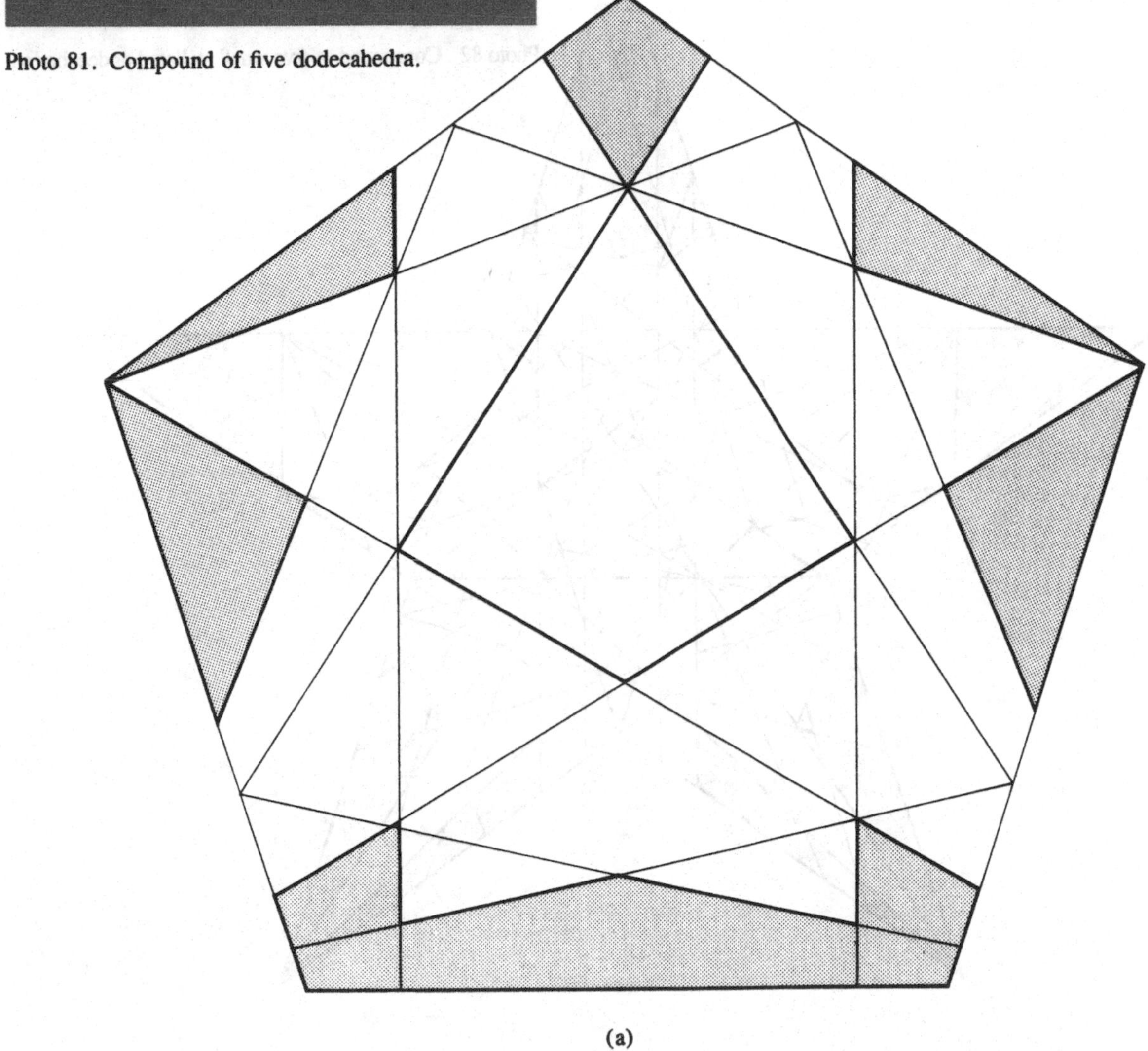

(a)

and 62 is worth careful investigation, because these were such fruitful sources for models of nonconvex uniform duals. Photo 82 shows a model for the compound of five small stellated dodecahedra, made from the stippled portions in Fig. 110b, and Photo 83 is the compound of five great stellated dodecahedra, whose exterior parts are the shaded portions. Suitable

Photo 82. Compound of five small stellated dodecahedra.

Fig. 110. (b) Stellation pattern: compounds of five dodecahedra. (c) and (d) Nets for construction of model: compound of five great stellated dodecahedra.

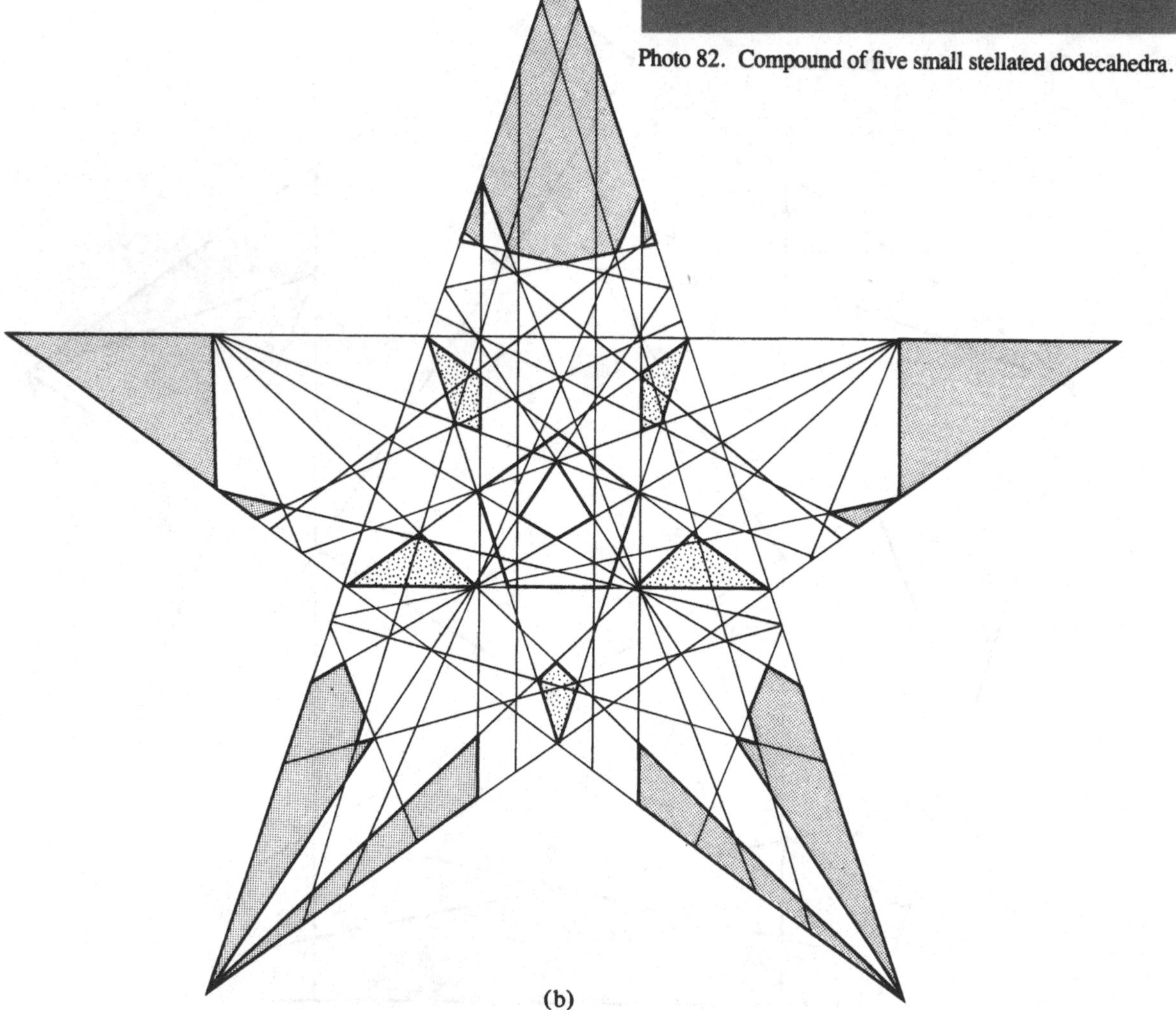

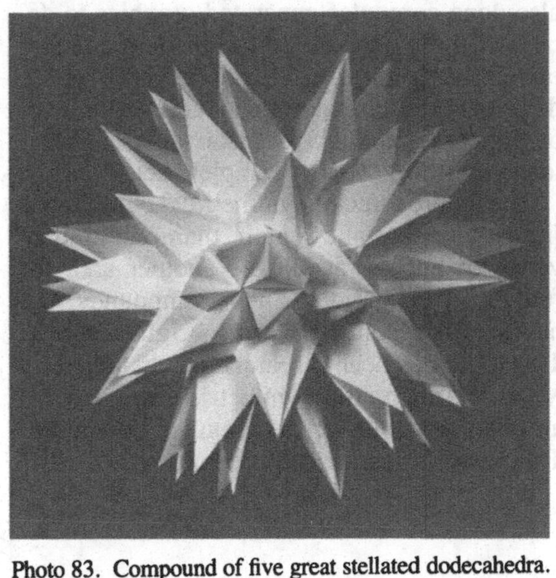

Photo 83. Compound of five great stellated dodecahedra.

nets for a model are shown in Fig. 110c, where the small wedge-shaped cells can be made by folding these portions and then letting the lower portions of these cells cover the shaded areas of the nets. These shaded areas may be left uncut.

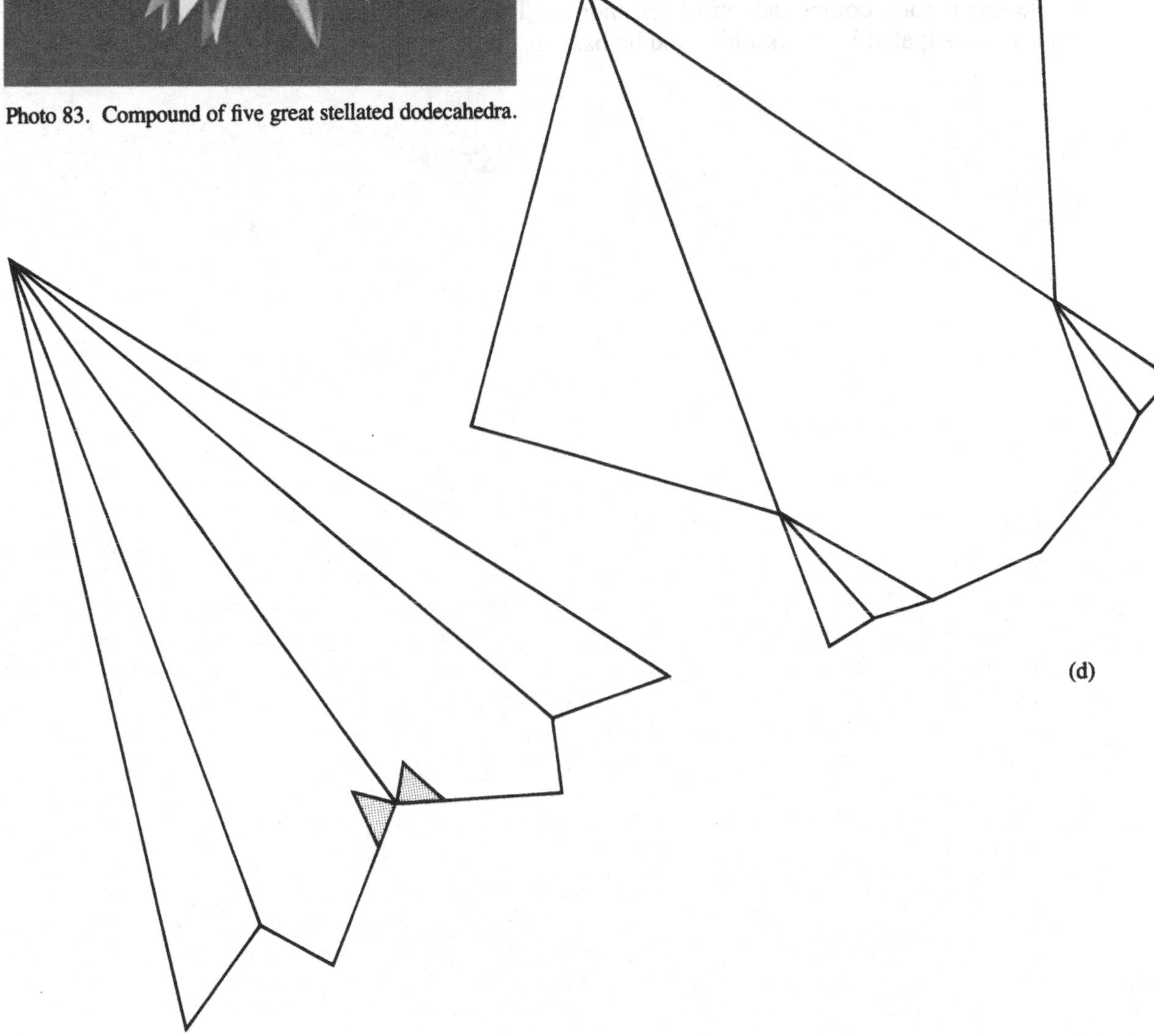

(c)

(d)

The compound of five great dodecahedra is also embedded in this pattern, but it needs considerable enlargement to show its exterior portions, broken as these are into numerous smaller cells along the edges. Thus a model is very complex in its structure. If you are really interested you may consult the list of references at the end of this book for further details.

The compound of two small stellated dodecahedra and the compound of five small stellated dodecahedra are examples of uniform compounds of uniform polyhedra. A uniform compound must have all its vertices alike. The whole set of such compounds has been thoroughly investigated by J. Skilling, and he has also done computer-generated graphics to depict all these shapes. Some of them are fantastically complex. I doubt that even I, dedicated model maker that I am, would ever be ambitious enough to make some of these, to say nothing about a complete set.

The whole topic of symmetrical compounds (i.e., those that allow more than one kind of vertex to appear) would be another area for further investigation. Such compounds can have regular solids as basic original forms. This topic has been researched and studied by M. G. Harman. But symmetric arrangements are also possible for semiregular polyhedra. The possibilities seem to be endless.

Epilogue

You may have noticed that none of the prisms or antiprisms have been presented in this book. These are also classified as uniform polyhedra, because for these, all the faces are composed of regular polygons, and all the vertices are alike. Why are they omitted? I would say merely because I personally do not find these shapes particularly attractive or interesting. But a complete presentation of duals for uniform polyhedra should really take these into consideration. The *n*-gonal prisms have duals that become *n*-gonal dipyramids, and the *n*-gonal antiprisms have duals that become *n*-gonal antidipyramids. You are invited to investigate their shapes by consulting the available literature on this topic. It certainly has its own mathematical value; so it is worth any attention you may want to give to it.

You may also want to go far more deeply into the mathematics of polyhedral shapes, more deeply than was done in this book. I have deliberately limited the level of mathematics so that it might come within the realm of knowledge attainable by most people. But the tools of higher mathematics can be used, and in fact have been used and undoubtedly will continue to be used in the study and investigation of polyhedra, not only in three-dimensional space but also in the study of polytopes in higher dimensions.

Another major tool for further study and investigation is computer analysis. The age of the computer has already dawned, and every area of human endeavor has already been affected by its appearance. Computer graphics, also in color, can produce vivid pictures to aid the imagination. Computer-generated motion pictures for showing continuous transformations of polyhedral shapes are well within the realm of computer capabilities. These are avenues down which future polyhedral research can proceed.

In this book, duality was approached through the stellation process. It can also be approached from the viewpoint of the faceting process. Just as stellation may be thought of as the addition of solid cells to a basic solid, so the faceting process may be thought of as the removal of basic cells from an original solid. The cells so removed are generated from the intersection of planes on the interior of the given solid, and these planes intersect along lines joining vertices of the given solid, so that these lines are also interior to the solid. I have always found the stellation process easier to work with when model making is involved. It is easier to add cells made by the same technique of construction as that used for any of the other models. It is more difficult to remove such cells in a physical way. But once a model has been constructed, it is easier to imagine what takes place in the faceting process. Thus, all polyhedra, even nonconvex stellated forms, and not only uniform polyhedra, have their duals, which can be derived from the faceting process. Here is another broad area for polyhedral research.

It remains for you to continue along any avenue or path on which your own attraction may lead you. I hope you will be sufficiently challenged by the material that has been presented in this book, and I know that any model or models you make will bring satisfaction to you as well as to others, who at least admire the beauty of these shapes, even though they may never want to make a model for themselves.

Appendix: Numerical data

Polyhedron	Face angles of dual of convex hull	Face angles of dual
Numerical values calculated by the author		
20	108 (5 times)	108 (5 times)
21	108 (5 times)	36 (5 times)
41	108 (5 times)	60 (3 times)
22	60 (3 times)	36 (5 times)
70	60 (3 times)	104.47751 (3 times) 135.52249 (3 times)
80	60 (3 times)	15.522481 (3 times) 224.47751 (3 times)
87	60 (3 times)	15.522481 (3 times) 135.52249 (3 times)
73	63.434949 (twice) 116.56505 (twice)	41.810316 (twice) 138.18968 (twice)
94	Same as **73**	63.434949 (twice) 116.56505 (twice)
69	81.578942 (3 times) 115.26317	16.842116 (twice) 81.578942 278.42106
86	Same as **69**	16.842116 (twice) 98.421058 (twice)
92	Same as **69**	16.842116 81.578942 (twice)
77	31.399715 (twice) 117.20057	31.399715 117.20057 (twice) 94.199145
85	Same as **77**	31.399715 (3 times) 265.80086
103	Same as **77**	31.399715 (twice) 62.79943 (twice)
81	30.480325 (twice) 119.03935	51.335802 98.183872 (twice) 112.29645
88	Same as **81**	51.335802 30.480325 (twice) 247.703355
101	Same as **81**	30.480325 (twice) 81.816125 (twice)
72	67.78301 86.974155 (twice) 118.26868	67.783010 25.242834 (twice) 241.73132

Polyhedron	Face angles of dual of convex hull	Face angles of dual
74	Same as **72**	25.242834 (twice) 91.025844 (twice)
97	Same as **72**	6.0516864 86.974156 (twice)
Numerical values supplied by G. Fleurent		
71, 82, 90	69.353522 86.825041 (twice) 116.996397	12.661079 50.342524 116.996397 142.318554
95	64.386424 88.016693 (twice) 119.580189	20.554443 138.891115 20.554443
104	Same as **71, 82, 90**	50.342524 79.314957 50.342524
75	55.105901 69.788198 55.105901	18.699407 142.601186 18.699407
76, 83, 96	56.308466 67.383068 56.308466	41.403622 58.184446 80.405932 141.003690
99, 105, 109	Same as **75**	18.699407 69.788198 91.512395 128.911209
79	41.409622 52.300725 86.289653	7.420695 41.409622 131.169683
93	39.236218 54.031689 86.732094	22.062191 51.407782 106.530027
84	34.344076 57.075228 88.580696	21.624634 53.130102 105.245264
98	33.557310 57.702705 88.739985	8.142571 76.118258 95.739171
108	31.646551 59.135471 89.217979	13.192999 71.594636 95.212365
110	55.875334 (twice) 68.243333	115.682268 (5 times) 141.588659
111	66.996614 118.087861 118.413832 112.153962 124.347731	56.827663 114.144405 140.739123 114.144405 114.144405

Polyhedron	Face angles of dual of convex hull	Face angles of dual
112	65.452111	50.958266
	135.536636	112.175128
	84.691478	139.658778
	118.783138	112.175128 (3 times)
	135.536636	
113	67.018494	133.966698
	104.441185	101.508325 (4 times)
	112.423649	
	117.737512	
	138.376160	
114	66.308624	3.990130
	103.709183	103.709182
	103.709181	135.117677
	117.681507	103.709182
	148.591505	103.709182
116	67.242424	121.431614
	91.703105	75.357903 (4 times)
	120.221110	
	117.285674	
	143.547687	
117	69.522493	104.857465
	92.810020	18.785634 (4 times)
	145.255496	
	116.384016	
	116.027975	
118	30.788908 (twice)	21.031989 (4 times)
	118.422184	105.159945
115, 119	68.607458	90 (4 times)
	86.853436 (twice)	128.172708 (twice)
	117.685671	

References

Brückner, M. *Vielecke und Vielfläche*. Teubner, 1900.

Catalan, E. Mémoire sur la théorie des polyèdres. *Journal de l'Ecole Polytechnique* XLI(1865):1–71.

Chilton, B. L. Principal shadows of the 12 pentagonal regular 4-dimensional objects (polytopes). *Leonardo* 13(1980):288-94.

Coxeter, H. S. M. *Regular polytopes*. Dover, 1973.

Coxeter, H. S. M. *Regular complex polytopes*. Cambridge University Press, 1974.

Cundy, H. M., and Rollett, A. P. *Mathematical models*. Oxford University Press, 1961.

Ede, J. D. Rhombic triacontahedra. *The Mathematical Gazette* XLII(1958):98–100.

Fleurent, G. M. Reflections on geometrical deduction of polyhedra. Unpublished manuscript, 1979.

Graziotti, U. A. *Polyhedra, the realm of geometric beauty*. San Francisco, 1962.

Harman, M. G. Symmetric compounds. Unpublished manuscript, 1974.

Holden, A. *Shapes, space and symmetry*. Columbia University Press, 1971.

Kerr, J. E., and Wetzel, J. E. Platonic divisions of space. *Mathematics Magazine* LI(1978):4.

Lalvani, H. *Transpolyhedra, dual transformations by explosion-implosion*. Lalvani, P.O. Box 1538, N.Y., N.Y. 10001. 1977.

Loeb, A. L. *Space structures, their harmony and counterpoint*. Addison-Wesley, 1976.

Pawley, G. Stellated triacontahedra. Unpublished manuscript, 1973.

Pugh, A. *Polyhedra, a visual approach*. University of California Press, 1976.

Skilling, J. The complete set of uniform polyhedra. *Philosophical Transactions of the Royal Society of London* 278(1975):111–35.

Skilling, J. Uniform compounds of uniform polyhedra. *Mathematical Proceedings of the Cambridge Philosophical Society* LXXIX(1976):447–68.

Smith, A. Some regular compounds of star-polyhedra. *The Mathematical Gazette*, LVII (1973):39–46.

Smith, A. Uniform compounds of the group U_4. *Proceedings of the Cambridge Philosophical Society* LXXV(1974):115–17.

Verheyen, H. *Dipolygonids, mobile generators of uniform polyhedra*. University of Antwerp, 1979.

Wenninger, M. J. *Polyhedron models*. Cambridge University Press, 1971.

Wenninger, M. J. *Spherical models*. Cambridge University Press, 1979.

Wenninger, M. J. Some interesting octahedral compounds. *The Mathematical Gazette* LII(1968):16–23.

Wenninger, M. J. Avenues for polyhedral research. *Structural Topology* (University of Montreal) 5(1980):5–26.

Wetzel, J. E. On the division of the plane by lines. *American Mathematical Monthly*, LXXXV (1978):8.

List of polyhedra and dual models

*Boldface numbers refer to model numbers in *Polyhedron models* (1971).

Printed in the United States
By Bookmasters